Preface

This book mainly deals with the social effects due to development and usage of energy at various levels. Energy shortages with rising cost confront every nation, especially those aspiring for industrial development. The economically developed countries would be able to tackle the problem by investing on technology to harness non – conventional energy sources. In addition to this there is a need to shift from energy sources emitting carbon dioxide to tackle global warming. As the world is facing energy crisis there is a need to explore the opportunities that help us to generate energy. This provides the context of this treatise.

Meeting the domestic energy demands of so many people is a challenge by itself. The Problem has been intensified as nations are aspiring for rapid economic development through industrialization. In their efforts to generate more energy, the countries are trying to tap non - renewable resources as well as the technologically newer ways to tap renewable sources. Social responsibilities for exploring, harnessing, adopting the given energy sources to obtain effective means to consume those are no less important. A time has come when no resource of this planet, including energy, should be used wastefully. To achieve this ideal, the society as a whole would require collective efforts of all groups, especially between the organizations harnessing energy and the consumers of energy. And both must understand the underlying principles of energy generation and energy consumption. The patterns of social responses to any and all types of energy resources have been referred. The issue of successful implementation of plans for energy harnessing is not purely a matter of technology. Social acceptance has come by attaining clearer understanding of the imperatives of a given energy source. There the technologists and the people must interact. While opting for the most sustainable energy use and to score success in that regard, every given country is conditioning their people to act as responsible consumers of energy. Where the societies at large have joined hands with the providers of energy by opting for processes to reduce waste and to attain efficiency in extracting the desired forms of energy like heat, light or sound, those countries are fairly successful in facing the energy crisis. One may not ignore the importance of responsible social actions in meeting energy needs. But the Process of collaboration has been complex. In nearly all countries, every choice made by the people on energy generation or consumption provoked controversies. Some contenders did not hesitate to use political platforms to score points. Judicial actions have been no less important in this regard. Political undercurrents not infrequently acted as intercepts on the processes of bringing together the visions of the technologists and the users of a given energy source. This aspect is dealt by including the values of responsible social behavior with regard to energy options

Dr Rakesh Namdeti

CONTENTS

List of Tables

List of Figures

Since1961. Biocapacity varies each year with ecosystem management, agricultural practices (such as fertilizer use and irrigation), ecosystem degradation, and weather.

Acronyms and Abbreviations:

kW   Kilo Watt
MW   Mega Watt
Wh   Watt hour
CFL   Compact Fluorescent Lamp
LED   Light Emitting Diode
IAEA   International Atomic Energy Agency
CEA   Central Electricity Authority
EPA   Environment Protection Agency
JNNSM   Jawaharlal Nehru National Solar Mission
HDI   Human Development Index
EAI   Energy Alternative India
HEEP   Home Energy Efficiency Program
GDP   Gross Domestic Product
GNP   Gross National Product
CFC   Chloroflurocarbons
UPCL   Uttarkhand Power Corporation Limited
QOL   Quality of Life

Conversions:

1. 1 Kilowatt Hour = 3,413 British Thermal Units (BTUs)
2. 1 Pound = 0.00045 Metric Tons
3. 1Btu (British thermal unit) = 778.14 ft lbf = 1.0551 kJ,
4. 1kJ = 0.94782 Btu= 25,037 lbmft/s^2
5. 1 Quad = 10^{15} Btu or $1.05*10^{15}$ kJ or $2.93*10^{11}$ kWh
6. 1 calorie = 4.18 J
7. 1 Therm = 10^{15} Btu
8. 1hp = 0.7064 Btu/s = 0.7457 kW = 745.7 W = 550 lbf ft/s = 42.41 Btu/min
9. 1barrel (42 gal) of crude oil = 5,800,000 Btu = 6120 MJ
10. 1 kWh = 859 kcal
11. 1 Joule = 1 Watt second
12. 1 toe = 42 GJ = 11.6 MWh
13. 1 eV = $1.6*10^{-19}$ J
14. 1 Hp = 746 W   1   Watt   =   668   Lumens

CHAPTER 1

This chapter deals with spacious perspective of energy with basic energy definitions and existence of energy in different forms. So in general definition of energy is ability to do work, but the energy can be defined in different ways in different scientific terms explaining with various examples. Even this chapter carries the information about the conversion of energies from one form to another form.

1.1 Energy Definition

Power: The rate of doing work is called power. Power is generated from nonrenewable energy sources such as coal, oil, and gas, or can be generated from natural resources such as geothermal, solar, and gravitational energies.

Work: Work is defined as the use of a force (F) to move something, where the force is the intensity with which we try to displace (push, pull, lift, kick, throw) an object. The amount of work to do depend on how much force is applied and what distance (d) is being covered.

Energy: Energy is the capacity of a physical system to perform work. It is expressed in joules. Energy exists in several forms such as heat, kinetic or mechanical energy, light, potential energy, electrical, or other forms. According to the law of conservation of energy, the total energy of a system remains constant, though energy may transform into another form. Two billiard balls colliding, for example, may come to rest, with the resulting energy becoming sound and perhaps a bit of heat at the point of collision.

People often use the term 'energy' without realizing that it is not material. Scientists define it as a kind of force. To help comprehension in this regard, consider the use we make of this force. We say that to do some work we need energy. So energy is the force needed to do any work. Therefore, we may be correct if we say that energy is needed to increase the total amount of work that is necessary to achieve economic development. Such expressions do not contradict the scientific definition of energy, but to understand energy from a scientific viewpoint, some more understanding is needed. The objective of this lecture is to provide a preliminary on the diverse ways that scientists have understood the features of energy.

Emergence of the concept:

In the early stages of development of science, energy was seen as a force to explain easily observable phenomena, such as in the effects detected in the properties of objects of any type of change therein. It uses to be argued that all changes in the properties of objects could in fact be explained as a result of application of some sort of energy. From this approach emerged the idea that objects can store energy, which can use in the future. As a result, it was suggested that there is some potential for changing as well as the actual change brought about. The stored amount is called the potential energy. The transformation is seen as an expression of the realized effects caused by energy in a kinetic state (that is, in motion or in a dynamic state). The potential and realized effects manifest themselves in many of different forms. *Electrical energy stored in a battery, chemical energy stored in food, or thermal energy of a water boiler, etc., are common example of potential energy. Similarly, kinetic energy can be seen as an expression of face as is noticeable in moving water, flowing air or expanding gas.*

The nature of energy, however, remained somewhat undefined until William Thomson (Lord Kelvin) compiled all these observations into the laws of thermodynamics, which aided in the repaid development of explanations of physical processes by using the concept of energy. It also led to a mathematical formulation of the concept of entropy (i.e., the spread of energy to create uniform distribution) by Ludwig Boatman, and to the broadening of the laws of radiant energy initially developed by Josef Stefan. Both are related to transformation of energy.

Definition of energy in different scientific disciplines:

The concept of energy and its transformation are useful in explaining natural phenomena. But the exact context of various natural phenomena and transformations differ in the various fields of enquiry of the different scientific sub disciplines, like Physics, Chemistry, Biology, Geology, etc. The concepts contained in the different scientific fields, offer opportunities to use energy to perform diverse types of work in society.

In Physics, energy is seen as the cause of change in the position of an object, i.e., as movement of an object over a distance. Expanding steam can push a piston to rotate a wheel, which can move a carriage attached to wheels. One could note the types of movement involved in shifting the position of the carriage. First is the expansion of water molecules in all directions that force the piston within a cylinder to move in a linear direction, which in turn causes rotational motion to the wheels that carry the carriage on a linear path. All these changes are linked to the potential energy contained in coal, which transforms to kinetic heat energy while burning. Hence energy is seen by the physicists as a force producing the ability to do work.

In Chemistry, atoms and molecules are the central concepts, which are made up of electrons and protons. Therefore some forces are seen at work during the rearrangement of atoms at the time of formation and transformation of molecules. This is what happens to the burning coal in the example above. Within the field of Chemistry, a transformation is possible only if the quantity of free energy decreases. Free energy, although difficult to define and can be seen as a synthesis of the concepts of energy and entropy. This is where physics and chemistry have a common interface. Entropy implies that energy from a concentrated source is consistently dissipated so as to be distributed over the universe to ultimately establish a uniform order. Hence from a chemical point of view, molecules are seen to settle down to a lower state of energy, thereby honoring the law of entropy. In the process free energy is released. This is a useful concept for all other scientific fields.

In Biology, the concept of energy refers to a complex set of chemical reactions. In Biology, growth, development and metabolism are some of the central considerations. These cannot be explained without invoking the concept of free energy. Living organisms survive because chemical bonds are constantly broking to permit exchange and release of energy. Energy is derived due to breakage of bonds in molecules of carbohydrates, fats and oils, etc. These molecules, along with oxygen, are the common stores of concentrated energy that sustain the biological processes. Nearly all energy transformations studied from sunlight thought photosynthesis. When a person ingests food, this stored energy is partially released, through digestion. The transformed molecules settle down in to a lower energy state and the free energy is used to do the necessary work. Thus the exchange of metabolic energy in biotic (i.e., biological) organisms also enhances the entropy of the universe.

In Meteorology, phenomena like wind, rain, hail, snow, lightning, tornadoes or hurricanes, are all seen as result of energy transformations sustained by sunlight on the earth. It has been estimated that the average total solar incoming radiation, commonly described as insolation, is about 1350 watts per square meters at the equator at midday, this is known as the solar constant. More energy is received in the tropics than is radiated, whereas more energy is radiated at the poles than is received. Hence climatic balance is maintained by a transfer of energy from the heated tropics to the cooler poles. This transfer of energy is the force that drives the winds and the ocean currents. Like biological possesses, all meteorological processes involve transformation of energy from one concentrated form to a less concentrated form, thereby honoring the law of entropy.

In Geology, Volcanoes, earthquakes, continental drift, formation of mountain ranges, crust or the continents, are phenomena that are seen as a result of energy transfer from the core to the Earth's crust. As a result, geological forces of lava emission that lead to continental accretion or subduction (i.e., drawing downwards) of one of two colliding plates, spreading of the sea floor, etc., account for 90% of the Earth's energy. The tectonic are driven by heat from the Earth's interior, which through dissipation once again reconfirms the law of entropy.

In Geomorphology, the physical processes of erosion (also called mass wasting), transportation and deposition are seen as a result of interaction between solar energy and gravity that drives water, ice and wind to act s agents of mass wasting. An estimated 23% of the total insolation is used to drive the hydrological cycle, which starts with the formation of moisture through evaporation, and successively goes through condensation, precipitation of moving glaciers or flowing rivers to the sea, to retune the amount of water initially withdrawn through evaporation. The hydrological cycle sustains many other processes of transformation of matter. When water vapor condenses to fall as rain, it dissolves small amounts of carbon dioxide and nitrogen oxide, making a weak acid. This acting Upton the metallic silicates that form most rocks, produces chemical weathering. Physical weathering of rocks is produced by the expansion of ice crystals left by water in the joint planes of rocks or by the rain drops. The products are transported by flowing water, moving glaciers, wind or waves. A geological cycle is continued when these eroded rocks are later uplifted into mountains or added to the contents to be subjected once again to the processes of mass wasting.

In Astronomy and Cosmology, stars, nova, supernova, quasars and gamma ray bursts lead to the universe's highest output of energy, due to all kinds of transformations of matter. All stellar phenomena (including of course solar activity) are driven by various kinds of energy transformations. Energy in such transformations can be due to the gravitational collapse of matter, but more usually is through transformation of molecular hydrogen into various classes of astronomical objects like stars black holes etc. A part of it is caused by nuclear fusion of hydrogen.

Socially perceived meaning of energy: There are both natural scientists and social scientists, in society. The distinction between natural scientists and the social scientists is important as their approach to energy is different. Then there are various groups of lay persons any society, who are also interested in energy. The scientific connotation of term energy seen by the natural scientists are not easy for other persons to understand. For most people, the term energy is synonymous with energy resources like fuels in furnaces and electrical power. When people talk about conservation energy in the context of the energy crisis, they mean on should use less energy. Energy in society has many interfaces, each one of which defines the contextual reference of the term.

In Economics, for example, energy is considered in terms of production and consumption of energy resources. All economic activities require energy resources. The ways in which a human society uses its existing energy resources develops means of producing or acquiring it, is a defining characteristic of its economy. The price of energy resources depends on the paying capacity of the consumers and the costs of its production at any given period of time. Studies on development of economics could include the progression from animal power to steam power, internal combustion engines and electricity, all of which are key elements in the development of modern civilization. These are the board areas where economists try to find equilibrium between demand and supply.

As the use of some energy resources also pollutes the ambient environment, environmental scientists, are concerned with the way energy is consumed. There are other sources of energy, such as solar power, tidal power, wind turbines, hydroelectric power, etc., which do not pollute the environment or do so in lesser amounts. The environmental sciences look upon these as a means of preserving a cleaner environment.

Exploration of energy resources is another aspect of the social context of energy. Scientist has realized that the known energy resources, such as fossil fuels, may not be sufficient for future needs. There is thus an urgent need to explore new deposits or well as never means of producing energy. Many scientists are exploring the possibility of cold fusion to tap the potential nuclear energy. Some countries are diverting significant economic resources towards space exploration with the exploration of discovering additional energy resources for our planet from other astral bodies. Research is under way to explore enzyme decomposition of biomass. Yet another field of research considers artificial photosynthesis, without the help of plants, as a future source of energy.

Since the cost of energy has become a significant factor in the economy of any society, management of energy resources has become very crucial. Energy management involves utilizing the available energy resources more effectively with minimum incremental costs. Reduction of energy consumption through the use of low energy appliances is one opportunity in this regard. Alternative uses of existing energy resources, like gas from coal is another field of management sciences. Many times it is possible to save expenditure on energy without incorporating fresh technology, by careful use of energy.

In the context of real politics, the owner ship and control of energy resources is playing an increasingly important role at the national and international level. at the national level, the given government seek to influence distribution of energy resources among various selections of society through pricing mechanisms or other administrative actions. at the international level, there may be transfer of national capital to posses the resources of the countries and thereby manipulate supply to sustain high prices that are unaffordable in poorer countries. Some political analysts maintain that the hidden reason for both the 1991 and 2003 wars can be traced to ensure strategic control of international energy resources.

While energy resources are essential ingredients for all modes of transport, the delivery of energy resources to the consumers is becoming equally important. Some energy resources like liquid or gaseous fuels are transported using tankers or pipelines, while delivery of electricity requires a network of grid cables. All these modes of transport of energy challenge scientist, engineers, policy makers, and economists, to make it more risk – free and efficient. These are subject areas of energy transport.

Energy Characteristics:

- ✓ Energy available in different forms.
- ✓ Energy can neither be created nor destroyed.
- ✓ Energy can be transformed from one form to the other form.
- ✓ Energy can be stored.
- ✓ Energy can be transported from one system to other or from one place to other.
- ✓ Free flow of energy takes place from higher potential to lower potential.

The forms of energy are graded as per their availability or energy. A high grade energy is completely convertible in to low grade energy on the other hand whole of low grade energy cannot be convertible in to work.

1.2 Different Forms of Energy

Energy has a number of different forms, all of which measure the ability of an object or system to do work on another object or system. In other words, there are different ways that an object or a system can possess energy. Here are the different basic forms:

Kinetic Energy: Consider a baseball flying through the air. The ball is said to have "kinetic energy" by virtue of the fact that it's in motion relative to the ground. You can see that it is has energy because it can do "work" on an object on the ground if it collides with it (either by pushing on it and/or damaging it during the collision). The formula for Kinetic energy, and for some of the other forms of energy described in this section will, is given in a later section of this primer.

Potential Energy: Potential gravitational energy, potential electric energy, potential magnetic energy, potential thermal energy, potential chemical energy, potential elastic energy and potential nuclear energy. Consider a book sitting on a table. The book is said to have "potential energy" because if it is nudged off, gravity will accelerate the book, giving the book kinetic energy. Because the Earth's gravity is necessary to create this kinetic energy, and because this gravity depends on the Earth being present, we say that the "Earth-book system" is what really possesses this potential energy, and that this energy is converted into kinetic energy as the book falls.

Thermal or heat energy: Consider a hot cup of coffee. The coffee is said to possess "thermal energy", or "heat energy" which is really the collective, microscopic, kinetic and potential energy of the molecules in the coffee (the molecules have kinetic energy because they are moving and vibrating, and they have potential energy due their mutual attraction for one another much the same way that the book and the Earth have potential energy because they attract each other). Temperature is really a measure of how much thermal energy something has. The higher the temperature, the faster the molecules are moving around and/or vibrating, i.e. the more kinetic and potential energy the molecules have.

Chemical Energy: Consider the ability of your body to do work. The glucose (blood sugar) in your body is said to have "chemical energy" because the glucose releases energy when chemically reacted (combusted) with oxygen. Your muscles use this energy to generate mechanical force and also heat. Chemical energy is really a form of microscopic potential energy, which exists because of the electric and magnetic forces of attraction exerted between the different parts of each molecule - the same attractive forces involved in thermal vibrations. These parts get rearranged in chemical reactions, releasing or adding to this potential energy.

Electrical Energy: All matter is made up of atoms, and atoms are made up of smaller particles, called protons (which have positive charge), neutrons (which have neutral charge), and electrons (which are negatively charged). Electrons orbit around the center, or nucleus, of atoms, just like the moon orbits the earth. The nucleus is made up of neutrons and protons. Some material, particularly metals, has certain electrons that are only loosely attached to their atoms. They can easily be made to move from one atom to another if an electric field is applied to them. When those electrons move among the atoms of matter, a current of electricity is created.

This is what happens in a piece of wire when an electric field, or voltage, is applied. The electrons pass from atom to atom, pushed by the electric field and by each other (they repel each other because like charges repel), thus creating the electrical current. The measure of how well something conducts electricity is called its conductivity, and the reciprocal of conductivity is called the resistance. Copper is used for many wires because it has a lower resistance than many other metals and is easy to use and obtain. Most of the wires in your house are made of copper. Some older homes still use aluminum wiring.

The energy is really transferred by the chain of repulsive interactions between the electrons down the wire not by the transfer of electrons per se. This is just like the way that water molecules can push on each other and transmit pressure (or force) through pipe carrying water. At points where a strong resistance is encountered, its harder for the electrons to flow - this creates a "back pressure" in a sense back to the source. This back pressure is what really transmits the energy from whatever is pushing the electrons through the wire. Of course, this applied "pressure" is the "voltage".

As the electrons move through a "resistor" in the circuit, they interact with the atoms in the resistor very strongly, causing the resistor to heat up - hence delivering energy in the form of heat. Or, if the electrons are moving instead through the wound coils of a motor, they instead create a magnetic field, which interacts with other magnets in the motor, and hence turns the motor. In this case the "back pressure" on the electrons, which is necessary for there to be a transfer of energy from the applied voltage to the motor's shaft, is created by the magnetic fields of the other magnets (back) acting on the electrons - a perfect push-pull arrangement!

Electrochemical Energy: Consider the energy stored in a battery. Like the example above involving blood sugar, the battery also stores energy in a chemical way. But electricity is also involved, so we say that the battery stores energy "electro-chemically". Another electron chemical device is a "fuel-cell".

Electromagnetic Energy (light): Consider the energy transmitted to the Earth from the Sun by light (or by any source of light). Light, which is also called "electro-magnetic radiation". Why the fancy term? Because light really can be thought of as oscillating, coupled electric and magnetic fields that travel freely through space (without there having to be charged particles of some kind around). It turns out that light may also be thought of as little packets of energy called photons (that is, as particles, instead of waves). The word "photon" derives from the word "photo", which means "light". Photons are created when electrons jump to lower energy levels in atoms, and absorbed when electrons jump to higher levels. Photons are also created when a charged particle, such as an electron or proton, is accelerated, as for example happens in a radio transmitter antenna.

But because light can also be described as waves, in addition to being a packet of energy, each photon also has a specific frequency and wavelength associated with it, which depends on how much energy the photon has (because of this weird duality - waves and particles at the same time - people sometimes call particles like photons). The lower the energy, the longer the wavelength and lower the frequency, and vice versa. The reason that sunlight can hurt your skin or your eyes is because it contains "ultraviolet light", which consists of high energy photons. These photons have short wavelength and high frequency, and pack enough energy in each photon to cause physical damage to your skin if they get past the outer layer of skin or the lens in your eye. Radio waves, and the radiant heat you feel at a distance from a campfire, for example, are also forms of electro-magnetic radiation, or light, except that they consist of low energy photons (long wavelength and high frequencies - in the infrared band and lower) that your eyes can't perceive. This was a great discovery of the nineteenth century - that radio waves, x-rays, and gamma-rays, are just forms of light, and that light is electro-magnetic waves

Sound Energy: Sound waves are compression waves associated with the potential and kinetic energy of air molecules. When an object moves quickly, for example the head of drum, it compresses the air nearby, giving that air potential energy. That air then expands, transforming the potential energy into kinetic energy (moving air). The moving air then pushes on and compresses other air, and so on down the chain. A nice way to think of sound waves is as "shimmering air".

Nuclear Energy: The Sun, nuclear reactors, and the interior of the Earth, all have "nuclear reactions" as the source of their energy, that is, reactions that involve changes in the structure of the nuclei of atoms. In the Sun, hydrogen nuclei fuse (combine) together to make helium nuclei, in a process called fusion, which releases energy. In a nuclear reactor, or in the interior of the Earth, Uranium nuclei (and certain other heavy elements in the Earth's interior) split apart, in a process called fission. If this didn't happen, the Earth's interior would have long gone cold! The energy released by fission and fusion is not just a product of the potential energy released by rearranging the nuclei. In fact, in cases, fusion or fission, some of the matter making up the nuclei is actually converted into energy. How can this be? The answer is that matter itself is a form of energy! This concept involves one of the most famous formulas in physics, the formula $$E = mc^2$$

This formula was discovered by Einstein as part of his "Theory of Special Relativity". In simple words, this formula means the energy intrinsically stored in a piece of matter at rest equals its mass times the speed of light squared.

When we plug numbers in this equation, we find that there is actually an incredibly huge amount of energy stored in even little pieces of matter (the speed of light squared is a very large number!). For example, it would cost more than a million rupees to buy the energy stored intrinsically stored in a rupee coin at our current (relatively cheap) electricity rates. To get some feeling for how much energy is really there; consider that nuclear weapons only release a small fraction of the "intrinsic" energy of their components.

Forms of energy and their relationship:

The examples given in the previous section suggest that energy binds all subjects that scientists are looking at through their respective disciplines. The community of scientists considers energy in several different forms, such as thermal, chemical, electrical, radiant, nuclear etc. these forms of energy, as stated earlier, are mainly classified into two broad groups, kinetic energy and potential energy. However, some forms of energy, cannot be classified in this manner. Heat is a varying mix of both potential and kinetic energy. Kinetic energy refers to the motion of a body or particles within it. Thermal energy (i.e., heat) comes from the motion of atoms or molecules within a solid, liquid or gaseous substance. Similarly, radiation energy is conceived as being caused by moving photons. However the phenomenon is more complex than what is observed commonly. Electrical energy is seen as the transfer of electrons from one place to another. But in reality, an electric current does not involve any net transfer of electrons from one ends of the conducting wire to another since the speed of electrons at both is the same. This means that the number of electrons that return is the same as the number emitted. This creates some complexity in the common man's understanding of the full meaning of kinetic energy in the context of electricity. The speed of electrons at both ends of a wire remaining the same means that their kinetic energy is the same and, therefore, cannot be the source of energy delivered to a load in between. This feature is better viewed as an attribute of electricity.

Radiation energy also cannot be neatly categorized as classical kinetic energy; since photons have no invariant mass and thus the energy required to accelerate their motion cannot be calculate by using other kinetic equations. In short, there still persists some ambiguity and some unexplained elements, in our attempts to clarify our understanding of energy.

Potential energy is due to the position of one object relative to other objects. it can either be the work done on an object by a force, or work done by the object against a force. Some examples indicate the intrinsic nature of the statement, but once again do not provide a full explanation of all forms of potential energy. It is better not to provide comprehensive explanation as each explanation mind the objective of this book. However it is useful to know the different ways in which potential energy can be observed, which are given below.

Potential gravitational energy: For example, is the work of gravitational force during rearrangement of mutual position of interacting masses, such as when masses are moved apart or brought closer together; say from accretion of interstellar mate stellar materials while forming our planet or from expulsion of expanded rock in molten state as lava.

Potential electric energy is the ability of electric forces to do work during rearrangement of positions of charges. This also includes some of the common chemical potential energies, that is to say, energy required to break chemical bonds, or obtained from forming them. The energy released in lightning and from using an amount of electrical power from an electrical wiring system, are all common examples of electromagnetic potential energy.

Potential magnetic energy can also remain stored in magnetic field such field are intrinsic properties of certain particles such as in magnets, but these often result from the relative motion of electric charges in an electrical current such as in dynamo. Magnetic potential energy is closely related to electric potential energy, since both types of potential are mediated by the electromagnetic field. An electrical transformer is an example of storage of magnetic potential energy.

Potential thermal energy is the energy stored in the equilibrium of the molecules. This potential energy is derived from 'deformation' of atomic bond during motions of atoms. as atoms oscillate around their position of equilibrium, they not only have kinetic energy of motion but also a potential energy of displacement from the equilibrium position.

Potential chemical energy is liberated when the bonds of chemical structures are rearranged. The mixture of a fuel and oxygen is an example. The energy exists as potential energy to be converted to heat when the material, seen commonly as fuel, combines with oxygen. Other common examples of chemical potential are a rechargeable battery, or food items.

Potential elastic energy is stored in the elastic nature of objects. Elastic energy is actually of several types. It is sometimes a kind of electric potential energy as in metal springs, where from energy is released as changed compressed atoms are allowed to move apart. However, in other cases, such as compressed air or rubber bands, the potential energy is stored in the arrangement of atoms, which can allow rapid conversion of thermal energy into work, when these are rearranged.

Potential nuclear energy along with electric potential energy, constitutes the energy released from nuclear fusion and nuclear fusion processes. In both cases strong nuclear force bind nuclear particles more strongly and closely, after the reaction has completed. Weak nuclear forces provide the potential energy for certain kinds of radioactive decay, such as beta decay. Nuclear particles like protons and neutrons are not destroyed in fission processes, but collections of them have less mass than if they were individually free. This mass difference is liberated as heat and radiation in nuclear reaction. The energy from the sun, also called solar energy, is an example of this form of energy conversion. In the sun, the process of hydrogen fusion converts parts of the solar matter into light, which is radiated into space.

Review Questions

1. Distinguish the definition of energy from scientific disciplines and socially perceived meaning of energy.

2. Define energy in scientific disciplines clearly bringing out the characteristics of it?

3. Define work, power and Energy and mention their standard units.

4. Do you agree the statement "The tides, winds and water are the sources formed from the Sun" and give your justification.

5. Define energy in engineering terms and mention its units?

6. Differentiate energy from power with relevant example.

7. What are the conversion energies involved in the following sources?

   a) Microphone b) Motor c) Generator d) Automobile Engine

   e) Water falling on the turbine f) A ball thrown into the air

   g) Electric Bulb   h) Bunsen burner   i) Projectile

8. Mention the different forms of energy and explain with examples.

Objective type Questions

1. Which of these energy changes in a flash light?
   a) Chemical to electrical   b) electrical to light
   c) Chemical to electrical to light to heat   d) light to heat

2. Which of these energy changes in a vehicle?
   a) fuel to electrical   b) Fuel to heat to mechanical
   c) Heat to mechanical   d) None of these

3. What is the unit of energy?
   a) Joules   b) Newton/Second
   c) ergs   d) Both a and c

4. Which among the following conversion energies exist when a ball thrown into the air
   a) Mechanical to K.E   b) P.E to Mechanical to K.E to P.E
   c) K.E to P.E   d) P.E to mechanical

5. The energy humans get from food originally comes from:

   a) Sugar   b) Meat   c) Vegetables   d) The Sun

6. Which of the following is not a Primary Energy Source?

   a) Oil   b) Natural Gas   c) Electricity   d) Wood

7. Energy available in fuels is stored as

   a) Heat Energy   b) Chemical Energy
   c) Atomic Energy   d) Explosive Energy

---

8. The ultimate energy source for the Earth is:

a) Electricity   b) Natural gas   c) The Sun   d) Plants

CHAPTER 2

ENERGY RESOURCES

This chapter deals with the classification of various energy resources with perspicuously bringing out the technical aspects of production of each resource and making comparison of energy resource with India and the rest of the world.

2.1 Classification of energy resources

Renewable energy (non-conventional): Renewable energy is energy which comes from natural resources such as sunlight, wind, rain, tides, and geothermal heat, which are naturally replenished. In 2011, about 31.95% of energy production came from renewable energy, with 21.53% coming from hydro, 10.42% from other renewable energy sources like solar, wind, biomass, geothermal and tidal waves. The share of energy production from sustainable source like nuclear energy in India is about 2.7% with having six nuclear power plants in operation with production of 4,780 MW whereas seven other nuclear reactors are under construction. India's ambitious plan to reach nuclear power energy of 63,000 MW in 2032.

Table 2.1 The current Installation capacity of different energy sources in India *(Source: Central Electricity Authority, 2011)*

Energy Source	Amount of Energy Production in MW	Percentage %

Coal	96,743	54.66
Gas	17,706	10
Oil	1,199	0.67
Hydro	37,367	21.53
Other renewable	18,454	10.42
Nuclear	4,780	2.7
All resources	1,76,990	100%

About 16% of global final energy consumption comes from renewable, with 10% coming from traditional biomass, which is mainly used for heating, and 3.4% from hydroelectricity. New renewables (small hydro, modern biomass, wind, solar, geothermal, and biofuels) accounted for another 2.8% and are growing very rapidly. The share of renewable in electricity generation is around 19%, with 16% of global electricity coming from hydroelectricity and 3% from new renewable.

While many renewable energy projects are large-scale, renewable technologies are also suited to rural and remote areas, where energy is often crucial in human development. Globally, an estimated 3 million households get power from small solar PV systems. Micro-hydro systems configured into village-scale or county-scale mini-grids serve many areas. More than 30 million rural households get lighting and cooking from biogas made in household-scale digesters. Biomass cook stoves are used by 160 million households.

Climate change concerns, coupled with high oil prices, peak oil, and increasing government support, are driving increasing renewable energy legislation, incentives and commercialization. New government spending, regulation and policies helped the industry weather the global financial crisis better than many other sectors.

Prospects of Energy Efficiency in Non Conventional power production:

Energy sources described as non - conventional in nature are solar power, wind power, tidal power, geothermal power and biofuels. These are also called renewable energy sources. Renewable energy is of great importance because the fossil fuel supplies will run out sooner. But looking into other country profiles on the order of energy harnessed from renewable sources, one may wonder whether the time has come to solely depends on this sources of energy for other future welfare. Without reducing the importance of developing these as alternative energy sources. it is wise to believe that they will not be viable for large-scale use in decades to come. For the present, they can be useful in meeting small scales power diamonds working in tender with the conventional power sources. Taking this as a remind, we may explore the options that that we have in India on extended use of different types of renewable energy sources one by one.

Solar Energy: The total solar energy flux intercepted by the earth on any particular day is 4.2×10^{18} Watt-hours or 1.5×10^{22} Joules (or 6.26×10^{20} Joules per hour). This is equivalent to burning 360 billion tons of oil (toe) per day or 15 Billion toe per hour.

There are two ways of looking at the prospect of using solar heat in India. In the first instance, we need to look into the technological base that can support prompt servicing of energy generating appliances. From this angle, it would be advantageous to popularize direct use of heat from sunshine without the support of a solar panel generating electricity to run a heater. The most common use would be to heat water to cut down electricity consumption for baths and save fuel in the kitchen. Such devices can be promoted for use in all parts of India even if the given location does not have a cloudless sky every day of the year. The idea use to lesson pressure on coal fired thermal power plants to the maximum extent possible. The second use of solar heat would be to produce steam with solar furnaces with the objective of increasing electricity production without emitting pollutants. The steam can be used to turn turbines and drive generators. Such drives would be most effective in and around the deserts of India where the sky remains cloudless on most days. By deserts we do not mean only the hot deserts of Rajasthan, but the cold desert of ladakh too, which would also support solar - furnaces equally effectively. Experimentation is needed to find the smallest effective size of such a plant.

Solar electricity: Solar energy is tapped using both solar thermal and Photovoltaic cell. In Solar thermal the solar energy is used to produce steam which is subsequently used to drive a turbo-generator to produce electricity. But in case of a Photovoltaic cell semiconducting materials are used to produce positive and negative charges under the influence of light i.e. Energy from sun light. Solar energy can alternatively, also be used for water heating, Air Conditioning and Cooking

Solar photovoltaic cells convert sunlight into electricity and photovoltaic production has been increasing by an average of more than 20 percent each year since 2002, making it a fast-growing energy technology. At the end of 2010, cumulative global photovoltaic (PV) installations surpassed 40 GW and PV power stations are popular in Germany and Spain. Many solar photovoltaic power stations have been built, mainly in Europe. As of October 2011, the largest photovoltaic (PV) power plants in the world are the Sarnia Photovoltaic Power Plant (Canada, 97 MW), Italy - 84.2 MW, Germany-80.7 MW, Ukraine-80 MW, Germany, 71.8 MW, Rovigo Photovoltaic Power Plant (Italy, 70 MW), Spain-60 MW, and Germany- 54 MW.

The first Indian solar thermal power project is in progress in Phalodi (Rajasthan). The solar thermal power plant has cost 4 times as much as the coal based steam thermal power plant. Some large projects have been proposed, and a 35,000 km^2 area of the Thar Desert has been set aside for solar power projects, sufficient to generate 700 to 2,100 GW.

Solar thermal power stations operate in the USA and Spain, and the largest of these is the 354 megawatt (MW) SEGS power plant in the Desert. The scope for extensive use of devices to produce electricity from sunshine is constrained by many factors in India. It is true that development of solar technology began forty or fifty years ago and admirable progress has been made, especially in the last two decades. A great number of solar technologies for the production of electricity are now available from commercial channels. Many of these devices have been installed on a significant scale in both developed and developing countries. They are used in different ways; either stand alone or incorporated in conventional energy networks and grids. They are already providing energy services to individual homes in villages and cities; however most of them should still be viewed as pilot projects. To move to country wide applications of solar technologies in villages, cities, islands and mountains, a lot of social support has to be organized. This will only come if the people are helped to better understand the infrastructure required solar energy, whether installed in remote areas or in cities. The targeted beneficiaries do not always know what the considerations are for them to go in for solar energy, and what is can be used for. Moreover, first time users would have to be assured that after sales service of reliable quality would be available. This would help in overcoming initial resistance arising from a feeling of uncertainly. However, this is exactly where the available capacity for servicing is inadequate.

In theory, a photovoltaic (PV) cell is very simple. However, one needs a solar panel to produce the desired amount of electricity, which requires adequate roof space with an appropriate slope facing the sun. This is not readily available in many Indian houses. Then the solar panel, otherwise sealed by rubber, can be ruined due to water seeping into it as the heat of the tropical regions turns the rubber brittle. Moreover, a solar cell's direct current will not run normal house hold appliance. It must be converted to alternating current with an inverter, which fail frequently and services to repair them are not available in remote areas. Even in cities, the available expertise is still poor. Quite often they need replacing, which is costly. At sites far from power grids, a battery bank is also needed to store electricity for times when the sun is not shining. With acid and zinc plates, batteries are not only heavy and expensive, but potentially dangerous too.

Even then many individuals is interested in installing solar electric systems for their homes but do not know that such a system takes from 10 to 16 years to pay back depending on the location. This is the reason for providing rebate programs and tax credits. In India these have to be given by the government and the amount of money required is high. The same amount, if than not, better results.

Under the circumstances, only remote areas would be suitable targets for solar electric systems supported by a battery bank. Such systems can also be installed in places served by electricity grids. However, in these locations, two actions are necessary. In the first instance, the homes should have 'smart meters' to allow them to sell the surplus power at certain times to feed the utility grids. Smart meters turn the meter backward when feeding power from the solar electric system to the utility companies. Secondly, the utility companies must be willing to buy electricity from the private unite when the power produced would not be needed for their own consumption. The price of purchased power should be at par with what they charge their clients. However, the policy trust should be on reducing the consumption of fuel wood.

Hydro electricity: Hydro-electric power, using the potential energy of rivers, now supplies 17.5% of the world's electricity (99% in Norway, 57% in Canada, 55% in Switzerland, 40% in Sweden, 7% in USA). Water is about 800 times denser than air and therefore even a small stream of water can produce a reasonable amount of energy. Water energy can be in many different forms such as hydroelectric energy, micro hydro, hydro power without dams, and ocean energy.

Today there are many hydroelectric power stations in the world and together they are providing about 20% electricity across the globe.

Wind farms: Grouping 10 to 100 wind turbines together in so called "wind farms" can lead to savings of 10% to 20% in construction, distribution and maintenance costs. India has gained a lot of experience in harvesting wind power. India is the fifth country in the ranking of energy production from wind. This has come primarily in the shape of wind farms connected to the regional and national power grids. The wind farms are of different sizes. There are a few isolated wind mills located in the remote areas, which are not connected to power grids. possibly the gain from wind power can be increased by developing wind mills in remote areas as well as in power starved rural tracts, which would be connected to local grids if convenient. The accent should be on using the power in kitchens to reduce consumption of fuel – wood. Wind power is growing at the rate of 30% annually, with a worldwide installed capacity of 158 GW in 2009, and is widely used in Europe, Asia, and the United States.

Tidal & Wave power: There is no instance of India attempting power generation from tides and waves. The high order of sedimentation along the India coast limits the development of effective sites. But illuminating the buoys by tapping wave movements can offer good dividends.

Hydrogen: It is promising energy sources. But the technology of production of fuel cell is generally absence in India. Production of membranes would be a formidable task. Production of hydrogen would be less problematic, if splitting water in to oxygen and hydrogen is chosen as the safest technology. It would be more attractive to generate hydrogen from hydrocarbon fuels. But storing and transporting hydrogen will be a problem. Potential users of fuel cells in India must be made aware of hydrogen, its utility and safely consideration.

Geothermal energy: Hardly any electrical power is produced in India from geothermal sources. Steam generation may not be immediately possible from geothermal power of reasons due to the high cost of drilling in to the crust to reach the heat reserves and setting up a power plant. But heat exchange between homes and the ground at a depth of 15 feet can be gainfully attempted in urban tracts to cut down costs of cooling and heating buildings. The world's largest geothermal power installation is The Geysers in California, with a rated capacity of 750 MW.

Biogas: As a fuel, biogas has been in use UN India over a long time. This gas is extracted from gobar gas plants, using cow dung and farm waste. Perhaps a change in the design of the plant can find application in wider regions. There are two basic components of these plants is a well some 30 feet deep. In many places in the valleys of the Ganges and the Brahmaputra Rivers, the ground water table comes close to the surfaces, which requires construction of a leak proof well, which is expensive. These wells surrounded by ground water cool down the biodegradation process, leading to sub optimal production of gas. Secondly, the store houses of the bio-gas are generally made of galvanized metal, which in a humid climate corrodes quickly, due to contact with the gas. Replacement of the store houses is time consuming and costly.

India may consider the utility of a device that is used for producing bio- gas in Taiwan, Vietnam and Laos. It consists of a 30 feet long plastic tube laid on gently sloping land. The biotic inputs take about 21 days to be completely digested. Water seal device in the input section prevents the gas from escaping. The input channel leads to the digester tube a little below the top of the tube. A container attached to the top of the tube traps the gas. The nutrient rich effluent and semi fluid container residues are released through the bottom of the tube. Both are useful materials for agricultural farming. The energy efficiency of the bio- gas can be enhanced by using other biota absorb CO_2 and thereby enrich the methane content of the residual gas.

Ethanol: The prospect of using ethanol to save petrol is bright in India as long as it is extracted from the fibrous residues of sugarcane and from molasses left as residue in sugar production. To try to produce ethanol directly from sugar may assure better outputs, but that would require increasing quantities of land to raise sugarcane. The demand of sugarcane is also very high. Sugarcane stands in the field for nearly 10 months and thereby prevents multiple cropping in a year. There are many contending demands on the available stock of usable water in India. Therefore, expending the area under sugarcane would imply reduction of the area under food crops in addition to causing water scarcity. The costs to compensate these losses would expectedly be very much higher than the savings on petrol from the use of ethanol.

Brazil has one of the largest renewable energy programs in the world, involving production of ethanol fuel from sugar cane, and ethanol now provides 18% of the country's automotive fuel. Ethanol fuel is also widely available in the USA.

Biodiesel: The use of biodiesel would certainly reduce consumption of diesel. The non- edible crops that yield oil to produce biodiesel can grow on many types of lands. If actual allocation of land for these crops is limited to land that is otherwise unsuitable for cultivation of food crops, then there will be little upset in the economy and a lot can be gained from the use of biodiesel. The problem is the administration in the control. There is no administration machinery that can monitor the program comprehensively. It is observed that nearly all state governments are eager to invite investments for cultivation of biodiesel yielding crops. But savings from the use of biodiesel would far less than those from the loss of farmland.

Now to mention some advantages and disadvantages of renewable energies for e.g. wind power, wind as we all know is free and can't be purchased anywhere, therefore wind power is very cost efficient. It produces no greenhouse gases or waste that could cause harm to the environment. Another advantage is that the land beneath the tower can still be used for farming since it doesn't take up a lot of space. Last but not the least wind power is a very good way of providing energy to remote areas. Even though wind is free and can't be purchased from anywhere this sometimes can be a disadvantage for example on days when there is no wind all one can do is hope and pray for some wind because you can't just simply go out and purchase it from your local hardware store. Another disadvantage is that these wind farms are usually near the coast, and near the coast means higher land prices. Some people feel that these tall towers destroy the beauty of the land. Some other disadvantages of these wind farms are they can affect television reception for people in close proximity to them, they can be noisy, they can kill birds, and lastly the wind cannot always be predicted

Advantages of Renewable energy:

- ✓ No danger of depletion. They recur in nature and are inexhaustible.
- ✓ No fuel cost and hence negligible running cost of power plant.
- ✓ More sites specific and used for local processing. There is no need for transmission and distribution.
- ✓ Low energy density and more or less no pollution and ecological balance problems.
- ✓ Simple in design and construction and are made of local materials local skills and by local people.
- ✓ Rural areas can be better served with.

Disadvantages of Renewable energy:

✓ Low energy density requires large size plant and increased cost.

✓ Low operating temperatures and low efficiency.

✓ User has to make a huge additional investment before delivering any benefit from it.

✓ Much of the construction material used for devices themselves are energy intensive.

✓ Low efficiency results in large heat rejections and hence thermal pollution.

✓ They occupies more land.

Non Renewable Energy: A non-renewable resource is a natural resource which cannot be produced, grown, generated, or used on a scale which can sustain its consumption rate, once used there is no more remaining. These resources often exist in a fixed amount and are consumed much faster than nature can create them. Fossil fuels (such as coal, petroleum and natural gas) and nuclear power (uranium) are examples.

Coal, oil and gas are called as fossil fuels formed by organic decomposition of buried dead organism. Coal is crushed to a fine dust and burnt to produce energy whereas energy from oil and gas can be burnt directly. Crude oil (called "petroleum") is easier to get out of the ground than coal, as it can flow along pipes. This also makes it cheaper to transport. Natural gas provides around 20% of the world's consumption of energy, and as well as being burnt in power stations, is used by many people to heat their homes. It is easy to transport along pipes, and gas power stations produce comparatively little pollution. Other fossil fuels are being investigated, such as bituminous sands and oil shale. The difficulty is that they need expensive processing before we can use them; however Canada has large reserves of 'tar sands' , which makes it economic for them to produce a great deal of energy this way.

Burning coal produces sulphur dioxide, an acidic gas that contributes to the formation of acid rain. This can be largely avoided using "flue gas desulphurization" to clean up the gases before they are released into the atmosphere. This method uses limestone, and produces gypsum for the building industry as a by-product. However, it uses a lot of limestone

As far as we know, there is still a lot of oil in the ground. But although oil wells are easy to tap when they're almost full, it's much more difficult to get the oil up later on when there's less oil down there. That's one reason why we're increasingly looking at these other fossil fuels.

Advantages

- ✓ Very large amounts of electricity can be generated in one place using coal, fairly cheaply.
- ✓ Transporting oil and gas to the power stations is easy.
- ✓ Gas-fired power stations are very efficient.

Disadvantages

- ✓ Basically, the main drawback of fossil fuels is pollution. Burning any fossil fuel produces carbon dioxide, which contributes to the "greenhouse effect".
- ✓ Burning coal produces more carbon dioxide than burning oil or   gas.
  It also produces sulphur dioxide, a gas that contributes to acid rain. We can reduce this before releasing the waste gases into the atmosphere.
- ✓ Mining coal can be difficult and dangerous. Strip mining destroys large areas of the landscape.

2.2 Modern Forms of Energy

The modern forms of energy are the resource which are clean and more sustainable to the environment, means no harm to environment. Such as the energy harnessing from Solar, Wind, Hydel, Biomass, Geothermal, Tidal and Hydrogen are called as modern forms of energy.

Solar Energy:

Solar technologies are broadly characterized as either active solar or passive solar depending on the way they capture, convert and distribute solar energy.

Active Solar energy can be captured in two forms, either as heat or as electrical energy.

- **Thermal Systems:** Thermal systems capture the Sun's heat energy in some form of solar collector and use it to mostly to provide hot water or for space heating, but the heat can also used to generate electricity by heating the working fluid in heat engine which in turn drives a generator.
- **Photovoltaic Systems:** Photovoltaic systems capture the sun's higher frequency radiation (visible and ultra violet) in an array of semiconductor, photovoltaic cells which convert the radiant energy directly into electricity.

Active solar techniques include the use of photovoltaic panels and solar thermal collectors to harness the energy. The harnessing of energy through active solar techniques either directly using photovoltaics (PV), or indirectly using concentrated solar power (CSP). Concentrated solar power systems use lenses or mirrors and tracking systems to focus a large area of sunlight into a small beam. PV solar panels have positive and negative layers. That means one side of each cell has an abundance of electrons, while the other has a shortage of them. This imbalance creates an electrical field in the cell, allowing the passing of electrons from the negative to the positive layer. Those electrons can be utilized for usage or it can be stored in the batteries for further use. This stored electrical energy then can be used at night. SPV can be used for a number of applications such as domestic lighting, street lighting village electrification water pumping desalination of salty water powering of remote telecommunication repeater stations and railway signals. Solar energy is the most readily available source of energy. It does not belong to anybody and is, therefore, free. In Solar thermal the solar energy is used to produce steam which is subsequently used to drive a turbo-generator to produce electricity.

Passive solar techniques include orienting a building to the Sun, selecting materials with favorable thermal mass or light dispersing properties, and designing spaces that naturally circulate air.

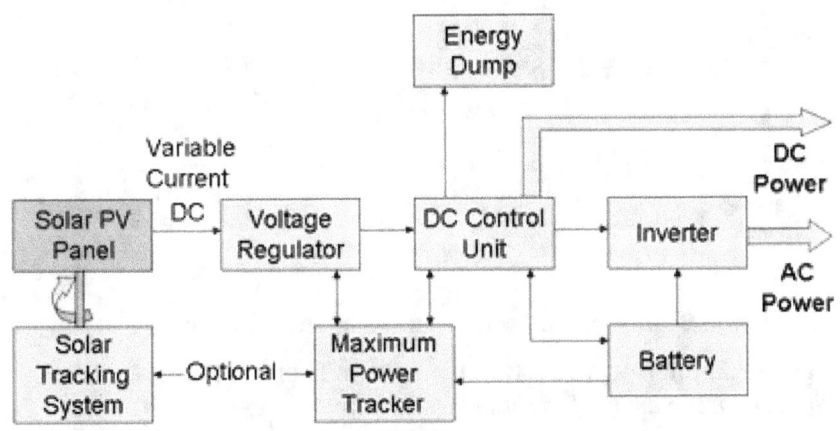

Photovoltaic Electric Power Generation

Figure 2.1: Energy production from Solar PV panels.

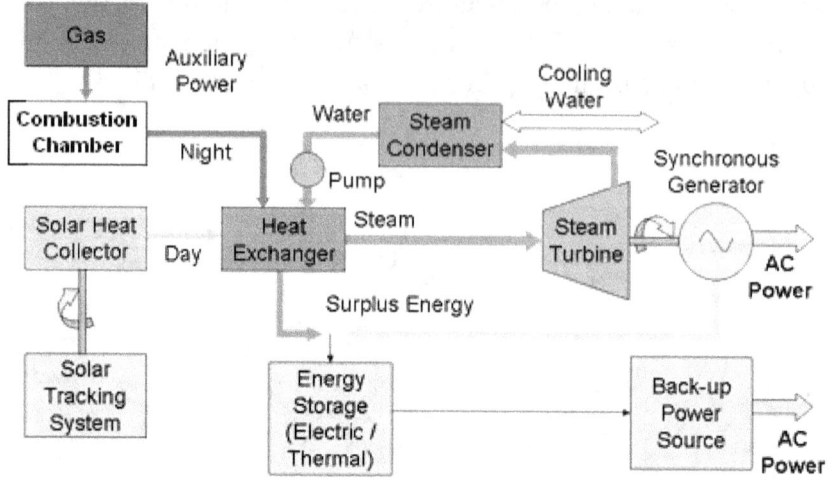

Large Scale Electric Power from Solar Thermal Energy

Figure 2.2: Energy production from Solar Thermal Plant
(Source: mpoweruk/electropaedia/energy sources)

For the environment, solar energy represents a cleaner, greener alternative to fossil fuels. Solar panels emit no carbon dioxide as they operate, helping the planet to minimize its carbon footprint.

In the next few years it is expected that millions of households in the world will be using solar energy as the trends in USA and Japan show. In India too, the Indian Renewable Energy Development Agency and the Ministry of Non-Conventional Energy Sources are formulating a program to have solar energy in more than a million households in the next few years.

India is one of the few countries with long days and plenty of sunshine, especially in the Thar desert region. This zone, having abundant solar energy available, is suitable for harnessing solar energy for a number of applications. In areas with similar intensity of solar radiation, solar energy could be easily harnessed. Solar thermal energy is being used in India for heating water for both industrial and domestic purposes. A 140 MW integrated solar power plant is to be set up in Jodhpur but the initial expense incurred is still very high. The Jawaharlal Nehru National Solar Mission (JNNSM) has set ambitious targets for power generation from solar energy in India. The Mission aims to have about 10 GW of grid-connected solar power plants by 2022.

Table 2.2 Planning for solar energy production by JNNSM

Phase 1	2010-2013	500 MW
Phase 2	2013-2017	1,500 MW
Phase 3	2017-2022	7,000 MW
Total		10,000MW

(Source: Government of India/Ministry of New and Renewable energy/JNNSM)

In addition to the JNNSM, several state governments have separate solar policies (Rajasthan, Gujarat and Karnataka) and many other state governments (Tamil Nadu, Maharashtra, Andhra Pradesh etc) are drafting solar policies on their own.

Solar photovoltaic cells convert sunlight into electricity and photovoltaic production has been increasing by an average of more than 20 percent each year since 2002, making it a fast-growing energy technology. At the end of 2010, cumulative global photovoltaic (PV) installations surpassed 40 GW and PV power stations are popular in Germany and Spain.

Many solar photovoltaic power stations have been built, mainly in Europe. As of October 2011, the largest photovoltaic (PV) power plants in the world are the Sarnia Photovoltaic Power Plant (Canada, 97 MW).

Biomass:

In this system Bagasse, Forestry and agro residue & Agricultural based industrial wastes are burnt to produce steam which is used to produce electricity. Biomass (plant material) is a renewable energy source because the energy it contains comes from the sun. Through the process of photosynthesis, plants capture the sun's energy. When the plants are burnt, they release the sun's energy they contain. In this way, biomass functions as a sort of natural battery for storing solar energy. As long as biomass is produced sustainably, with only as much used as is grown, the battery will last indefinitely.

The bio energy is harnessed by four ways and is given below table.

Table 2.3 Bio energy production with different resources

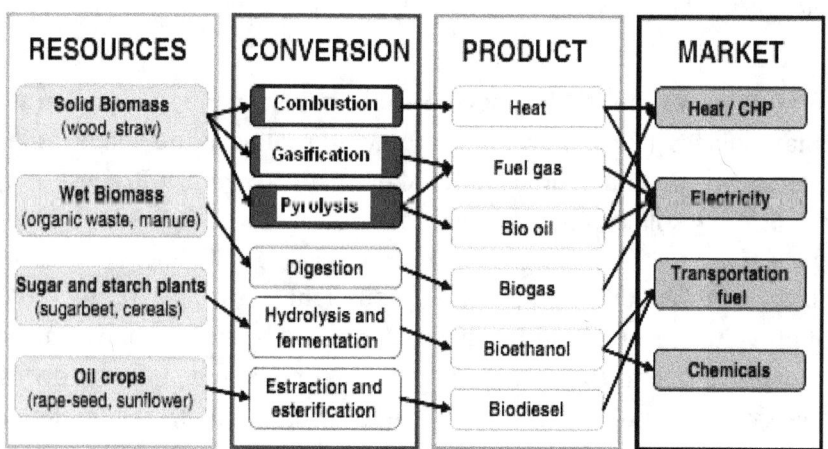

(Source: European Biomass Industry Association, Elsevier Applied Sciences Publishers ltd and Risoe National Laboratory)

India is very rich in biomass and has a potential of 16,900 MW (agro-residues and plantations), 5000 MW (bagasse cogeneration) and 2700 MW (energy recovery from waste). Biomass power generation in India is an industry that attracts investments of over INR 600 crores every year, generating more than 5000 MW of electricity and yearly employment of more than 10 million man-days in the rural areas.

Bio-energy contribution to the total primary energy consumption in India is over 27%. Indeed, this is the case for many other countries, because biomass is used in a significant way in rural areas in many countries. However, the contribution of biomass to power production is much smaller than this - currently, biomass comprises only about 2650 MW of installed capacity, out of a total of about 172000 MW of total electricity installed capacity in the country (May 2011).

India is the pioneer in biomass gasification based power production. While gasification as a technology has been prevalent elsewhere in the world, India pioneered the use of biomass gasification for power production. As a result, prominent Indian solution providers in biomass gasification are implementing their solutions in other parts of the world.

EAI(Energy Alternatives India) estimates the total installed capacity of biomass gasification based power production in India will be about 140 MW, out of a total of about 2600 MW of biomass based power (cumulative of grid connected and off grid). Of the total, bagasse based power generation has the lion's share (about 1400 MW), followed by combustion-based biomass power production (about 875 MW). While biomass gasification currently contributes little to power production, EAI foresees significant growth for this sector in future.

Geothermal Energy:
Geothermal energy is the earth's natural heat available inside the earth. This thermal energy contained in the rock and fluid that filled up fractures and pores in the earth's crust can profitably be used for various purposes. Heat from the Earth, or geothermal Geo (Earth) + thermal (heat) energy can be and is accessed by drilling water or

steam wells in a process similar to drilling for oil. Geothermal energy is an enormous, underused heat and power resource that is clean (emits little or no greenhouse gases), reliable (average system availability of 95%), and home grown (making us less dependent on foreign oil).

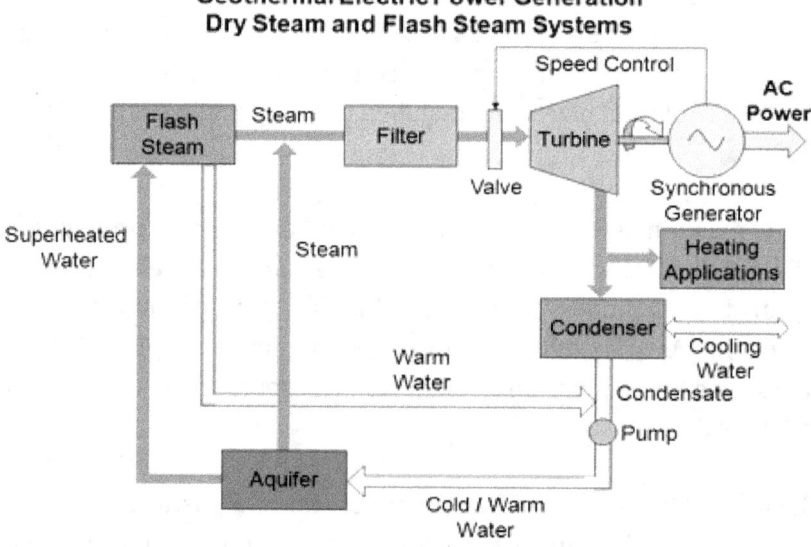

Figure2.3: Energy production from Geothermal
(Source of this figure: mpoweruk/electropaedia/energy sources)

India has reasonably good potential for geothermal; the potential geothermal provinces can produce 10,600 MW of power. But yet geothermal power projects has not been exploited at all, owing to a variety of reasons, the chief being the availability of plentiful coal at cheap costs. However, with increasing environmental problems with coal based projects, India will need to start depending on clean and eco-friendly energy sources in future; one of which could be geothermal.

Wind energy:

The Sun heats our atmosphere unevenly, so some patches become warmer than others. These warm patches of air rise, other air blows in to replace them - and we feel a wind blowing. We can use the energy in the wind by building a tall tower, with a large propellor on the top. The wind blows the propellor round, which turns a generator to produce electricity. Wind turbines operate on a simple principle. The energy in the wind turns two or three propeller-like blades around a rotor. The rotor is connected to the main shaft, which spins a generator to create electricity.

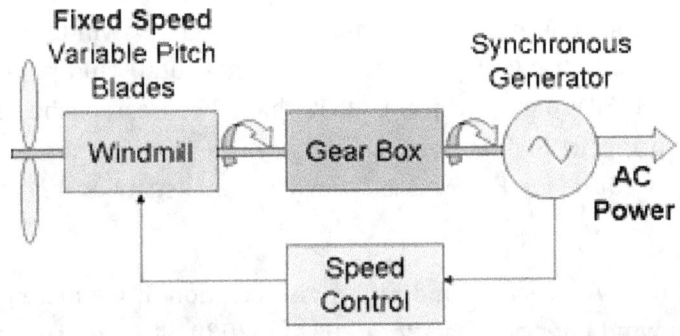

Large Scale Wind Power (Grid Systems)

Figure 2.4 Energy production from Wind *(Source: mpoweruk/electropaedia/energy sources)*

The best places for wind farms are in coastal areas, at the tops of rounded hills, open plains and gaps in mountains - places where the wind is strong and reliable. Some are offshore. To be worthwhile, you need an average wind speed of around 8 m/s. The propellors are large, to extract energy from the largest possible volume of air. The blades can be angled to "fine" or "coarse" pitch, to cope with varying wind speeds, and the generator and propellor can turn to face the wind wherever it comes from. Some designs use vertical turbines, which don't need to be turned to face the wind. The towers are tall, to get the propellors as high as possible, up to where the wind is stronger. This means that the land beneath can still be used for farming. Good wind sites are often located in remote locations, far from cities where the electricity is needed. Transmission lines must be built to bring the electricity from the wind farm to the city.

The development of wind power in India began in the 1990s, and has significantly increased in the last few years. As of June 2010, in India the installed capacity of wind energy was 12000 MW and was estimated to be 18000 MW by 2012. For India to reach 40 GW of renewable energy target in this decade, there is a compulsive need to have at least 25 GW of wind energy. The slow market adoption of other renewable energies opens up an attractive window of opportunities. India ranks 5th in the world with a total wind power capacity if 1080MW out of which 1025Mwhaev been established in commercial projects. Five nations – Germany, USA, Denmark, Spain and India – account for 80% of the world's installed wind energy capacity.

According to the American Wind Energy Association, if we increase our nation's wind energy capacity to 20% by 2030, it would reduce Greenhouse Gas Emissions a cumulative total of 7,600 million tons of CO_2 would be avoided by 2030, and more than 15,000 million tons of CO_2 would be avoided by 2050.

Hydro energy:

The Sun evaporates water from the sea and lakes, which forms clouds and falls as rain in the mountains, keeping the dam supplied with water. For free gravitational potential energy is stored in the water above the dam. Because of the great height of the water, it will arrive at the turbines at high pressure. The turbine turns a shaft that rotates a series of magnets past copper coils in a generator to create electricity. The water then returns to the river. From the powerhouse, transmission lines carry electricity to communities.

Hydro-electric power stations can produce a great deal of power very cheaply. Notice that the dam is much thicker at the bottom than at the top, because the pressure of the water increases with depth. Although there are many suitable sites around the world, hydro-electric dams are very expensive to build. However, once the station is built, the water comes free of charge, and there is no waste or pollution.In mountainous countries such as Switzerland and New Zealand, hydro-electric power provides more than half of the country's energy needs.

The present installed capacity as on June 2011 is approximately 37,367 MW which is 21.53% of total Electricity Generation in India.

Tidal Wave Energy

The ebb and flow of the tides can be used to turn a turbine, or it can be used to push air through a pipe, which then turns a turbine. The high energy of sea tides is used to rotate turbines which drive generators to produce electricity. The identified economic tidal power potential in India is of the order of 8000-9000 MW with about 7000 MW in the Gulf of Cambay about 1200 MW in the Gulf of Kutch and less than 100 MW in Sundarbans.

2.3 Technical Aspects of Energy

A power plant is basically an industrial facility for the generation of electrical energy. There are different types of power plants such as thermal, Hydel and thermal on mega scales, and geo-thermal, wind and solar on smaller scales. Here in this section it deals with only large scale units of electricity generation. The word 'generation' is a misnomer, because energy can neither be created nor destroyed; it can only be transformed. In power plants, energy is transformed from potential, mechanical or chemical type to electrical type. The processes of transformation are seldom efficient, and involve significant environmental impacts. However, as our lives get entwined more and more with the use of electrical equipments, we have to live with the inefficiencies.

Fossil fuel based Power Plant:

Thermal power plants produce and meet 40% of global electricity demand. Fossil fuels, such as natural gas, crude oil and its derivatives, coal and bio-mass can be mined, transported, stored and consumed with relative ease and lower cost. The process technology is simple and requires low-to-medium skilled workforce. Central Electricity Authority (CEA) data, as on 2011 June, 54.66 % or 96,743 MW of total electricity generation in India is coal-based.

Coal is abundant in nature. Anthracite, Bituminous and Lignite are coal types in the descending order of purity and grade. Coal mining and beneficiation are established industries. National Security worries are non-existent unlike nuclear material. For generations, households have used coal in their kitchen stoves and backyard barbeque grills.

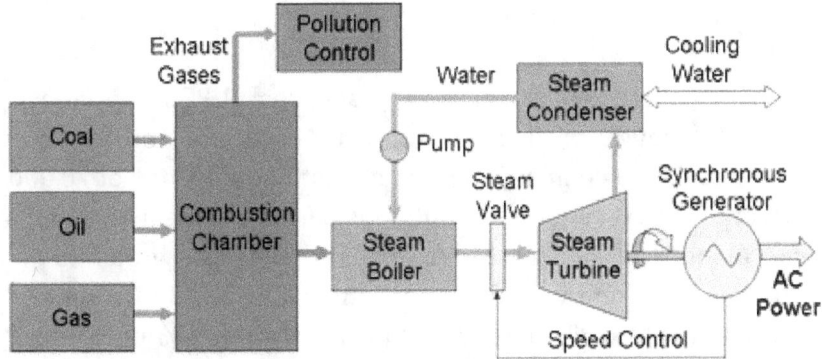

Fossil Fuel Powered Steam Turbine Electricity Generation
Figure 2.5 Illustration of Energy generation from fossil fuels
(Source of this figure: mpoweruk/electropaedia/energy sources)

The process of electricity generation begins with coal being fired in the boiler. This is a combustion reaction involving carbon in the coal and oxygen from the air. The products of the chemical reaction are flue gas and heat. The flue gas is composed of oxides of carbon, nitrogen and sulfur and particulate matter. This stage 1 is similar to what happens in our coal grills and kitchen stoves, but for scale and size of operations.

In the stage 2, the heat in the flue gas is transferred to feed water. It boils into steam. The steam pressure temperature is controlled in such a way that it can turn wheels of Steam Turbine and sustain high speeds. Steam pressures in the range of 200bar and at temperatures 540-600 deg C are typical in conventional thermal plants. Under such conditions, less than one-third of heat is recovered, remaining lost to the environment.

In the stage 3, the high speed turbines rotate generator rotors. Typically 6000 rpm with Faraday's principle of electromagnetic induction that mechanical work is converted into electrical energy. The used steam is condensed into water and recycled. The condenser must operate at lowest temperature like 30 degrees C and vacuum, and therefore requires cooling water is usually drawn from a river, lake or sea.

India has limited sources of high grade coal. UPCL is the first independent power producer in the country to use 100 per cent imported coal. The company imports almost one ship load of 70,000 tons every week. It requires 12,000 tons per day to fire a 1,000 MW power plant. The public tender specifies the coal grade to be bituminous and from Australia, South Africa, China and Indonesia. It restricts sulfur to 0.8% max and ash content to 4.75% max. These are stringent specifications, and if enforced and complied by the suppliers, in a tight and escalating coal price market, the company is doing what it says it is committed to do.

Hydel Power Plant: Hydroelectric power now supplies about 715,000 megawatts or 19% of world electricity. Large dams are still being designed. The worlds largest is the Three Gorges Dam on the third longest river in the world, the Yangtze River. Hydroelectric power can be far less expensive than electricity generated from fossil fuels or nuclear energy. Areas with abundant hydroelectric power attract industry. Environmental concerns about the effects of reservoirs may prohibit development of economic hydropower sources.

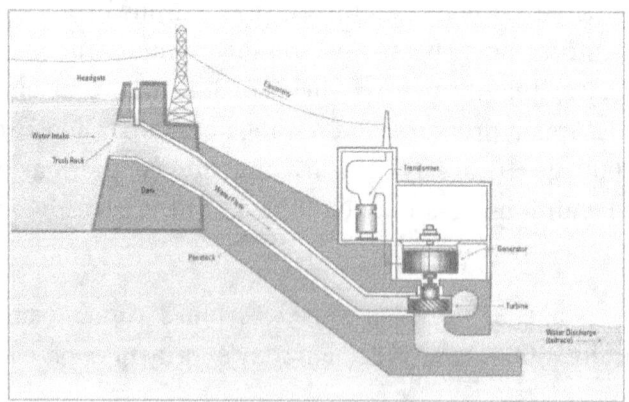

Figure 2.6: Electricity Production from hydro
(Source of this figure: Ontario power generation website)

According to the power generated, small hydro power is classified into small, mini/micro and pico hydro. In India, it is being classified as follows.

Small hydro	- 2 MW - 30 MW
Mini	- 100 kW - 2 MW
Micro	- 10 kW - 100 kW
Pico hydro	- 1 kW - 10 kW

The dam is constructed across a river, lake or any other water body. Water from the catchment area collects at the back of the dam to form reservoir. A pressure tunnel is taken off from the reservoir and water brought to valve house at the start of penstock.

The major components of a hydroelectric dam are as follows:

Dam: A barrier built across a watercourse to hold back the flow of water and create a reservoir. The reservoir that is formed is, in effect, stored energy.

Penstock: A pipeline used to convey water, under pressure, from the reservoir to the turbines of a hydropower plant.

Turbine: A machine that is turned by the force of the fast moving water pushing against its blades. Turbines convert the kinetic energy of the water to mechanical energy.

Generator: Connects to the turbine and rotates to produce the electrical energy.

Transformer: Converts electricity from the generator to usable voltage levels.

Transmission Lines: Conduct electricity from the hydropower plant to the electric distribution system. Transmission line voltages are normally 115 kilovolt or larger.

Nuclear Energy: Nuclear power is the use of sustained nuclear fission to generate heat and electricity. Nuclear power plants provide about 13–14% of the world's electricity, with the U.S, France and Japan together accounting for about 50% of nuclear generated electricity. In 2007, the IAEA reported there were 439 nuclear power reactors in operation in the world. As of 2010, India has 20 nuclear reactors in operation in six nuclear power plants, generating 4,780 MW while seven other reactors are under construction and are expected to generate an additional 5,300 MW.

Nuclear fission occurs when a neutron collides with a nucleus of a large atom such as Uranium and is absorbed into it causing the nucleus to become unstable and thus split into two smaller more stable atoms with the release of more neutrons and a considerable amount of energy. Naturally occurring isotope in which fission can be induced with thermal neutrons is Uranium-235 which splits into Barium-141, a soft silvery metal, and Krypton-92, an inert gas, and surplus free neutrons averaging about 2.4 neutrons per event. The process can be represented by the following equation:

$$^{235}U_{92} + {}^{1}n_0 \Rightarrow {}^{140}Ba_{56} + {}^{96}Kr_{36} + 3\,{}^{1}n_0 + 202MeV$$

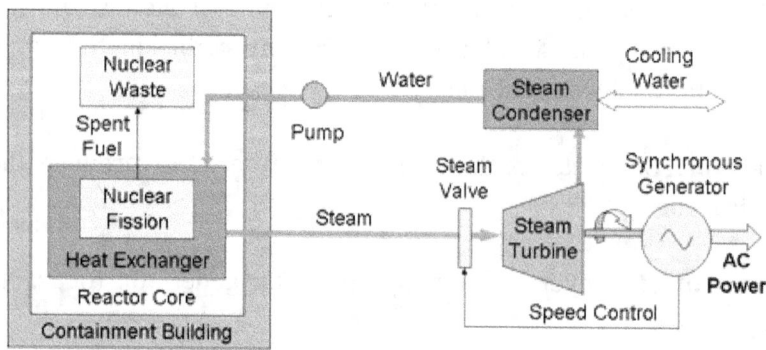

Boiling Water Reactor (Single Stage Heat Transfer)

Electricity Generation by Nuclear Power

Figure 2.7 Illustration of Nuclear energy production
(Source: mpoweruk/electropaedia/energy sources)

Coolants: The reactor core acts as a heat exchanger in which the coolant, which may be either a liquid or a gas, surrounds the fuel rods and captures the heat generated by the nuclear reaction. The coolant also acts as the thermal working fluid which is used either directly or indirectly to raise steam to drive a turbine generator.

Coolants must be good conductors of heat with low susceptibility to induced radioactivity and capable of operating at high temperatures. A variety of substances, including light water, heavy water, air, Carbon dioxide, Helium, molten metals such as Sodium, Sodium-Potassium alloy, Lead and Lead-Bismuth alloy as well as hydrocarbons (oils), have been used for this purpose.

Moderators:

The energy of the free neutrons must be within certain limits for fission to occur. High energy neutrons emitted by the fission process move too quickly to be captured by the fissile atoms and so must be slowed down or *moderated* to increase their chances of causing fission. Water, heavy water and graphite are moderators which are commonly used in the reactor core to slow down the neutrons. Certain hydrides, hydrocarbons, beryllium and beryllium oxide are also used for this purpose.

Control Rods:

A major safety system in nuclear reactors is provided by control rods of Boron, Cadmium or Graphite which absorb neutrons created by the fission process removing them from the active mass thus preventing further fissions from taking place. Because of their atomic structure these elements absorb neutrons, but do not fission or split. The rate of the chain reaction can be controlled by progressively inserting the control rods into, or removing them from the reactor core and the reactor can be quickly shut down by dropping the control rods into the core.

Review Questions

1. From the past 10 years we mainly depends our daily activities on electrical and electronic gadgets, that energy can be drew out from various resources. Classify those resources and enumerate advantages and disadvantages?

2. Demonstrate the production of energy from solar with neat sketch? Compare the current status of solar energy production in India and in the world.

3. State the prospects of different renewable power generation for a country like India?

4. What do you understand by the term direct conversion technologies and explain the power generation through these technologies?

5. What are the major types of electrical power plants? Briefly explain technical aspects of power plants existing in India?

6. What is geothermal energy? Sketch and explain the production of energy from geothermal?

7. With neat block diagrams explain generation of electrical energy from various Renewable and Non-Renewable resources?

8. Renewable source of energy are more appealing and acquiring outstanding role in current panorama. Mention the reasons why these resources are more importance?

9. Can we arrange solar P.V panels on roof of our buildings to meet our electricity requirement? Discuss pros and cons?

10. Demonstrate and depict the production of energy from Hydro and Thermal power plants?

11. What is nuclear fission and explain the energy production from nuclear fuel? Mention the current status of generation in India.

12. With a neat sketch, describe how a typical fossil fuel based power plant works?

13. How the energy can be harnessed from wind mills and explain its status of energy generation in India and the world?

14. Write about the share of different energy resources in India and the future of the availability of these resources.

15. Differentiate renewable and non renewable energy resources status in India and the world?

16. Demonstrate the production of energy from solar with neat sketch? Compare the current status of solar energy production in India and in the world.

17. Do you believe that it is a greater priority for wealthier, industrialized countries to install grid-tie PV systems or for poorer, less developed countries to adopt off-grid PV systems?

18. Discuss the advantages and disadvantages of renewable energy power plants.

Objective type Questions

1. World's largest producer of energy from Wind?
   a) Germany   b) China   c) USA   d) Italy
2. India's ranking position in world for the production of wind energy
   a) Second   b) Fourth   c) Fifth   b) Sixth
3. World's largest geothermal power plant is located at
   a) India   b) California   c) Brazil   d) Germany
4. The current percentage installation capacity of energy from coal in India is
   a) 80%   b) 54%   c) 25%   d) 90%
5. Renewable energy resources include all of the following except
   a) The sun.   b) The wind.   c) Biomass   d) Natural gas.
   e) Water
6. Which statement is true?
   a. Hydroelectric power, a nonrenewable resource, is generated from moving water
   b. Oil, a nonrenewable resource, is processed to make jet fuel
   c. Propane, a renewable resource, is used in some heating systems
   d. The energy from the sun, a nonrenewable resource, can be converted to electricity in solar cells
7. Which among the following is direction energy conversion technology?
   a) Solar photo voltaic b) Solar Thermal c) Wind d) Hydro electric
8. Which among the following energy production methods involve no noise pollution?
   a) Hydro electric   b) Wind   c) Thermal   d) solar photo voltaic
9. Biomass energy can be obtained from

(a) Energy plantations   (b) Petro crops

(c) Agricultural and urban waste biomass   (d) All of these.

10. Nuclear energy can be generated by
    (a) Nuclear fusion   (b) Nuclear fission
    (c) Both of these   (d) None of these

11. Identify the nonrenewable source of energy from the following
    (a) Coal   (b) Fuel cells   (c) Wind power   (d) Wave power

12. Which of the following is a disadvantage of most of the renewable energy sources?
    (a) Highly polluting   (b) High waste disposal cost
    (c) Unreliable supply   (d) High running cost

13. Which among the following resource cannot be regenerated?
    a. Solar   b. Biomass   c. Jatropha plants   d. Petrol

14. Which country contains about 50% of the world's coal resources?
    a) United States   b) China   c) Canada   d) the former soviet union

15. How many nuclear Power plants are there in India?
    a) Three   b) Four   c) Five   d) Six

16. How many Nuclear reactors are under working in India?
    a) 15   b) 20   c) 25   d) 30

17. How much percentage accounts for Hydro electricity generation in India as of 2011?
    a) 15%   b) 21%   c) 35%   d) 10%

18. Hydro electricity is
    a) Derived from Sun   b) Conversion efficiency is around 80 to 90%
    c) a renewable source   d) all of the above

19. Which of the following is not a Primary Energy Source?
    a) Oil   b) Natural Gas   c) Electricity   d) Wood

20. The major source of electrical power generation in India is
    a) Thermal   b) Hydel   c) Nuclear   d) Wind

21. Fossil fuels are made:
    a) From skeletons   b) In factories
    c) Over millions of years   d) Over a barbecue

22. A non-renewable energy resource:

a) Can be plugged in and recharged   b) Will eventually run out
c) Will not work for new appliances   d) Can be used over again

23. Which of these energy resources does not originate from the Sun's energy?
    a) Natural gas   b) Oil   c) Geothermal energy   d) Coal

24. At the moment renewable energy resources generate:
    a) More energy than fossil fuels   b) The same amount of energy
                                                            as fossil fuels
    c) Less energy than fossil fuels   c) Virtually no energy

25. Which of these statements can be used to describe renewable energy?
    a) They will run out   b) They pollute the environment
    c) Most renewable energy resources do not need burning
    d) all use the power Sun

26. Geothermal energy is:
    a) Generated in special flasks   b) Generated in rocks below the
                                            Earth's surface
    c) Generated by the Sun heating up rocks
    d) Generated by wearing thermal underwear

27. Nuclear fuel is:
    a) Non-renewable   b) Renewable   c) Burnt
    d) Environmentally friendly

28. The energy which is not derived from the sun is _____
    a) bio-mass   b) fossil fuels   c) nuclear energy d) geo-thermal

29. A solar cell converts _____
    a) Heat energy into electrical energy
    b) Solar energy into electrical energy
    c) Heat energy into light energy   d) solar energy into light energy

30. The source of energy of the sun is _____
    a) Nuclear fission   b) chemical reaction   c) nuclear fusion
    d) Photoelectric effect

31. Nuclear fusion reactions happens spontaneously in _____

a) The core of the earth   b) The commercial nuclear reactor
c) The atmosphere of the sun   d) The eruption of a volcano

32. Pick up the odd plant from the following:

a) Steam   b) Wind turbine generator plant   c) Tidal   d) Solar

33. Pick up the odd plant from the following

a) Steam   b) Hydro   c) Tidal   d) Nuclear Plant

34. Pick up the odd plant from the following

a) Thermal power plant   b) Nuclear   c) Geothermal
d) Solar PV panel

35. Heavy water in a nuclear reactor is used for

a) Cooling & moderation   b) Cooling and shielding radiation
c) Moderation & shielding radiation   d) To initiate reaction

36. Which is a conventional source of energy?

a) Solar   b) radio-active substances c) geothermal d) wind

37. The disadvantages of renewable source of energy is/are?

a) Intermittency   b) lack of dependability
c) Availability in low energy densities   d) All the above

CHAPTER 3

ECONOMICAL ANALYSIS

Our primary energy sources in this world are Renewable and Non-Renewable energy resources to meet our electricity demand. So in this chapter it shows the economical costs of all these resources. This will lucidly brings us the cost of distribution through several energy sources and also some costs are not included in the price, those costs can be explained in Hidden costs of energy.

3.1 Economical Aspects of Energy

Solar Power: The primary reason for the high cost of solar power is the high capital cost. Currently, it costs about Rs 15 crores per MW of capital cost for a solar PV power plant (MNRE has taken a benchmark capex of Rs 16 crores). About 50% of this cost is owing to the cost of the panels and the rest are for balance of systems (inverters, transformers, monitoring systems, wires and cables etc), the civil support infrastructure and the cost of installation. Among the balance of systems, inverters contribute the highest cost component, at about Rs 2.5 crores per MW.

The good news is, the cost of solar panels are showing dramatic decreases year on year. The bad news is, the balance of systems are not. So, solar (both PV and thermal) is 3-4 times as costly as wind.

As per UK data, the average solar radiation received is about 2.5 kWh / M² / day. A single solar panel with an area of 3 M² will therefore capture 2.5 x 3 x 365 = 2737 kWh of energy per year. With a system conversion efficiency of around 40% and less than optimal orientation of a typical rooftop mounted solar panel, the maximum usable energy received by a single panel system will be around 1000 kWh. The cost saving will depend on whether the solar system is replacing 1000 kWh of heating energy supplied by a gas or an electric water heating system and the associated tarriff charged for the energy. With UK domestic gas currently costing less than £0.03 per kWh ($0. 045) and electricity costing about £0.10 per kWh ($0.15) the annual savings are likely to be somewhere between £30 and £100 ($50 to $150). Since typical single panel installations cost around £2,500 or £3,000 ($4,000 to $5,000), unless the systems qualify for a government subsidy or there is a very large increase in energy costs, the payback time for the investment will be measured in decades rather than years. Saving the planet can be quite expensive.

Future trend expected of Solar Cell Prices:

PV solar panels come in mono crystalline, poly crystalline, amorphous and thin film varieties. Currently, crystalline silicon panels are the most commonly used PV systems. However, silicon is expensive and in short supply. High production costs make silicon panels too expensive for the average consumer in India. Companies are researching alternatives such as thin film systems as well as other non silicon options to bring down the cost of production and make PV solar panels widely accessible. Photovoltaic systems are developed using specific combinations of solar cells. Basically, PV systems are available in two designs, which are flat plate and concentrator panels. As of now, all PV solar panels sold in India are constituted of crystalline silicon cells.

Before widespread use, however, improvements are needed in the photovoltaic cells to make them economically competitive. Test photovoltaic cells that consist of silicon solar cells are currently up to 21% efficient in converting sunlight into electricity (Moore 1992). The durability of photovoltaic cells, which is now approximately 20 years, needs to be lengthened and current production costs reduced about fivefold to make them economically feasible. With a major research investment, all of these goals appear possible to achieve.

Currently, production of electricity from photovoltaic cells costs approximately Rs 30/kWh, but the price is projected to fall to approximately Rs 10/kWh by the end of the decade and perhaps reach as low as Rs 4 by the year 2030, provided the needed improvements are made. In order to make photovoltaic cells truly competitive, the target cost for modules would have to be approximately Rs 8/kWh. Locating the photovoltaic cells on the roofs of homes, industries, and other buildings would reduce the need for additional land by approximately 5%, as well as reduce the costs of energy transmission. However, photovoltaic systems require backup with conventional electrical systems, because they function only during daylight hours. Relating to the cost of solar cells in India, it is worth keeping in mind the cost per unit of solar power generation is Rs 15-17, much higher than coal-based electricity generation cost, in the range of Rs 4-9. Adoption of thin-film technology, organic cells and Concentrated Solar Power would go a long way in stabilizing solar cells prices in India to a competitive level.

Price of solar cells in India will fall if manufacturers of solar cells rise to the occasion. Solar energy is a long-term investment; the same applies to solar cells costs in India. Solar cells prices in India at the retail level are okay, but at factory level, the cost of production is high. Through continued investment and technical innovation, solar cells costs in India will gradually become competitive at the global stage.

Solar Ponds:

Solar ponds are used to capture solar radiation and store it at temperatures of nearly 100°C. Natural or man-made ponds can be made into solar ponds by creating a salt-concentration gradient made up of layers of increasing concentrations of salt. These layers prevent natural convection from occurring in the pond and enable heat collected from solar radiation to be trapped in the bottom brine.

The hot brine from the bottom of the pond is piped out for generating electricity. The steam from the hot brine turns freon into a pressurized vapor, which drives a Rankine engine. This engine was designed specifically for converting low-grade heat into electricity. At present, solar ponds are being used in Israel to generate electricity. In several locations in the United States solar ponds are now being used successfully to generate heat directly. The heat energy from the pond can be used to produce processed steam for heating at a cost of only Rs 2 to 3.5 per kWh Solar ponds are most effectively employed in the Southwest and Mid-west.

Solar Receiver Systems:

Other solar thermal technologies that concentrate solar radiation for large scale energy production include distributed and central receivers. Distributed receiver technologies use rows of parabolic troughs to focus sunlight on a central-pipe receiver that runs above the troughs. Pressurized water and other fluids are heated in the pipe and are used to generate steam to drive a turbogenerator for electricity production or provide industry with heat energy. The land requirements for the central receiver technology are approximately 1100 ha to produce 1 billion kWh/yr assuming peak efficiency, and favorable sunlight conditions like those in the western United States. Solar thermal receivers are estimated to produce electricity at approximately Rs 10 per kWh, but this cost is expected to be reduced somewhat in the future, making the technology more competitive Central receiver systems are being tested in Italy, France, Spain, Japan, and the United States (at the 10-megawatt Solar One pilot plant near Barstow, California). Also, Luz's Solar Electric Generating System plants at Barstow use distributed receivers to generate almost 300 MW of commercial electricity

Biomass energy systems:

At present, forest biomass energy, harvested from natural forests, provides an estimated 3.6 quads (1.1 x 10 18 Joules). Worldwide, and especially in developing countries, biomass energy is more widely used than in the United States. The cost of producing a kilowatt of electricity from woody biomass ranges from Rs 7 to 10 which is competitive for electricity production that presently has a cost ranging from 3 to 13. Approximately 3 kcal of thermal energy is required to produce 1 kcal of electricity. In these systems, the energy and economic costs would be significant and therefore might limit the use of this strategy.

Biomass will continue to be a valuable renewable energy resource in the future, but its expansion will be greatly limited. Its use conflicts with the needs of agricultural and forestry production and contributes to major environmental problems.

Ethanol:

A wide variety of starch and sugar crops, food processing wastes, and woody materials have been evaluated as raw materials for ethanol production. The total fossil energy expended to produce 1 liter of ethanol from corn is 10,200 kcal, but note that 1 liter of ethanol has an energy value of only 5130 kcal. Thus, there is an energy imbalance causing a net energy loss. Any benefits from ethanol production, including the corn by-products, are negated by the environmental pollution costs incurred from ethanol production. During the fermentation process approximately 13 liters of sewage effluent is produced and placed in the sewage system for each liter of ethanol produced.

Ethanol produced from corn clearly is not a renewable energy source. Its production adds to the depletion of agricultural resources and raises ethical questions at a time when food supplies must increase to meet the basic needs of the rapidly growing world population.

Methanol:

Methanol is another potential fuel for internal combustion engines. Various raw materials can be used for methanol production, including natural gas, coal, wood, and municipal solid wastes. At present, the primary source of methanol is natural gas. The major limitation in using biomass for methanol production is the enormous quantities needed for a plant with suitable economies of scale. A suitably large methanol plant would require at least 1250 tons of dry biomass per day for processing. Biomass generally is not available in such enormous quantities from extensive forests and at acceptable prices

Hydrogen:

Gaseous hydrogen, produced by the electrolysis of water, is another alternative to petroleum fuels. Using solar electric technologies for its production, hydrogen has the potential to serve as a renewable gaseous and liquid fuel for transportation vehicles. In addition hydrogen can be used as an energy storage system for electrical solar energy technologies, like photovoltaics.

At present, commercial hydrogen is more expensive than gasoline. For example, assuming Rs5 per kWh of electricity from a conventional power plant, hydrogen would cost Rs 9 per kWh. Therefore, hydrogen fuel may eventually be competitive. Some of the oxygen gas produced during the electrolysis of water can be used to offset the cost of hydrogen. Also the oxygen can be combined with hydrogen in a fuel cell, like those used in the manned space flights. Hydrogen fuel cells used in rural and suburban areas as electricity sources could help decentralize the power grid, allowing central power facilities to decrease output, save transmission costs, and make mass-produced, economical energy available to industry.

Wind Energy:

A typical domestic installation with a 1.75m swept diameter, (swept area of 2.4m^2), costs around £1500 ($2250). At the rated wind speed of 12.5m/s (28 mph) the wind power intercepted will be 2870 Watts, but after taking into account all the unavoidable system losses, the actual electrical output power will be around 1000 Watts. However this is at the upper end of the performance possibilities. Wind turbulence and shielding due to buildings and trees inhibits sustained strong, gust free wind flow and in any case, for most of the time, the wind speed will more likely be towards the lower end of the performance specification at 4 m/s (9 mph), that is a light breeze. At this speed the power output of the system will be about 32 Watts - Not enough to power a single light bulb. For much of the time the power generated could be less than the quiescent power drain of the inverter.

Running with a constant power output of 32 Watts for a full year would generate only 280 kWh (280 Units) of electrical energy worth £28 at today price of £0.10 ($0.15) per kWh. To put it into perspective, a typical UK household consumes about 5,000 kWh of electrical energy per year.

Coal-based and natural gas-based power cost anywhere between Rs 2 and 3 per KWh.

The economics of rural and remote locations make wind power more attractive than for urban locations. Because of the remoteness, connection to the electricity grid may be impossible or prohibitively expensive. Furthermore, larger, more efficient wind power installations are possible and the prevailing winds will also be higher.

Comparing the economics of different forms of electricity generation:

It is important to distinguish between the economics of nuclear plants already in operation and those at the planning stage. Once capital investment costs are effectively "sunk", existing plants operate at very low costs and are effectively "cash machines". Their operations and maintenance (O&M) and fuel costs (including used fuel management) are, along with hydropower plants, at the low end of the spectrum and make them very suitable as base-load power suppliers. This is irrespective of whether the investment costs are amortized or depreciated in corporate financial accounts assuming the forward or marginal costs of operation are below the power price, the plant will operate.

Capital costs comprise several things: the bare plant cost (usually identified as engineering-procurement-construction-EPC-cost), the owner's costs (land, cooling infrastructure, administration and associated buildings, site works, switchyards, project management, licences, etc), cost escalation and inflation. Owner's costs may include transmission infrastructure. The term "overnight capital cost" is often used, meaning EPC plus owners' costs and excluding financing, escalation due to increased material and labour costs, and inflation. Construction cost sometimes called "all-in cost", adds to overnight cost any escalation and interest during construction and up to the start of construction. It is expressed in the same units as overnight cost and is useful for identifying the total cost of construction and for determining the effects of construction delays. In general the construction costs of nuclear power plants are significantly higher than for coal- or gas-fired plants because of the need to use special materials, and to incorporate sophisticated safety features and back-up control equipment. These contribute much of the nuclear generation cost, but once the plant is built the cost variables are minor.

Long construction periods will push up financing costs, and in the past they have done so spectacularly. In Asia construction times have tended to be shorter, for instance the new-generation 1300 MWe Japanese reactors which began operating in 1996 and 1997 were built in a little over four years, and 48 to 54 months is typical projection for plants today.

Decommissioning costs are about 9-15% of the initial capital cost of a nuclear power plant. But when discounted, they contribute only a few percent to the investment cost and even less to the generation cost. In the USA they account for 0.1-0.2 cent/kWh, which is no more than 5% of the cost of the electricity produced.

Financing costs will depend on the rate of interest on debt, the debt-equity ratio, and if it is regulated, how the capital costs are recovered. There must also be an allowance for a rate of return on equity, which is risk capital.

Operating costs include operating and maintenance (O&M) plus fuel. Fuel cost figures include used fuel management and final waste disposal. These costs, while usually external for other technologies, are internal for nuclear power (ie they have to be paid or set aside securely by the utility generating the power, and the cost passed on to the customer in the actual tariff).

This "back-end" of the fuel cycle, including used fuel storage or disposal in a waste repository, contributes up to 10% of the overall costs per kWh, - rather less if there is direct disposal of used fuel rather than reprocessing. The $26 billion US used fuel program is funded by a 0.1 cent/kWh levy.

Calculations of relative generating costs are made using levelised costs, meaning average costs of producing electricity including capital, finance, owner's costs on site, fuel and operation over a plant's lifetime, with provision for decommissioning and waste disposal.

It is important to note that capital cost figures quoted by reactor vendors, or which are general and not site-specific, will usually just be for EPC costs. This is because owner's costs will vary hugely, most of all according to whether a plant is Greenfield or at an established site, perhaps replacing an old plant.

Nuclear plant: projected electricity costs (c/kWh)

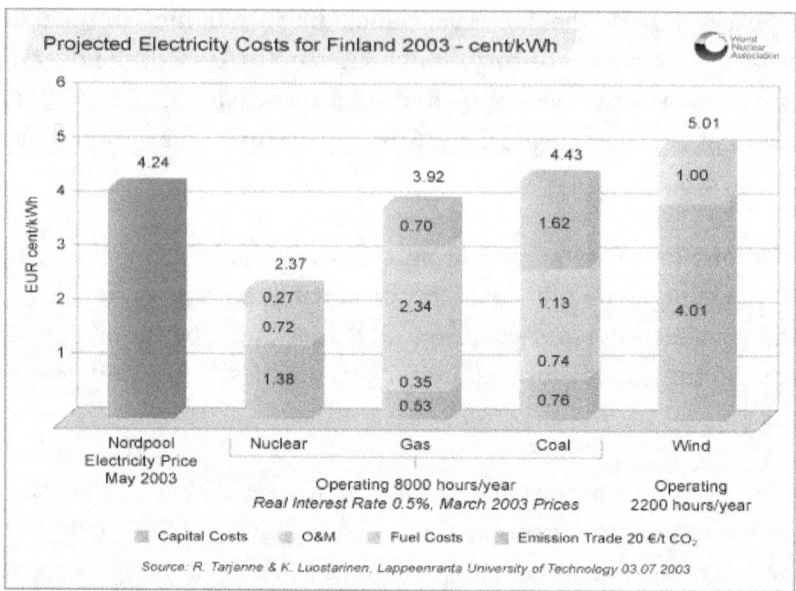

Figure 3.1 Projected Electricity costs

In the middle three bars of this graph the relative effects of capital and fuel costs can be clearly seen. The relatively high capital cost of nuclear power means that financing cost and time taken in construction are critical, relative to gas and even coal. But the fuel cost is very much lower, and so once a plant is built its cost of production is very much more predictable than for gas or even coal. The impact of adding a cost or carbon emissions can also be seen.

Electricity Generating Costs per kWh for Different Fuels:

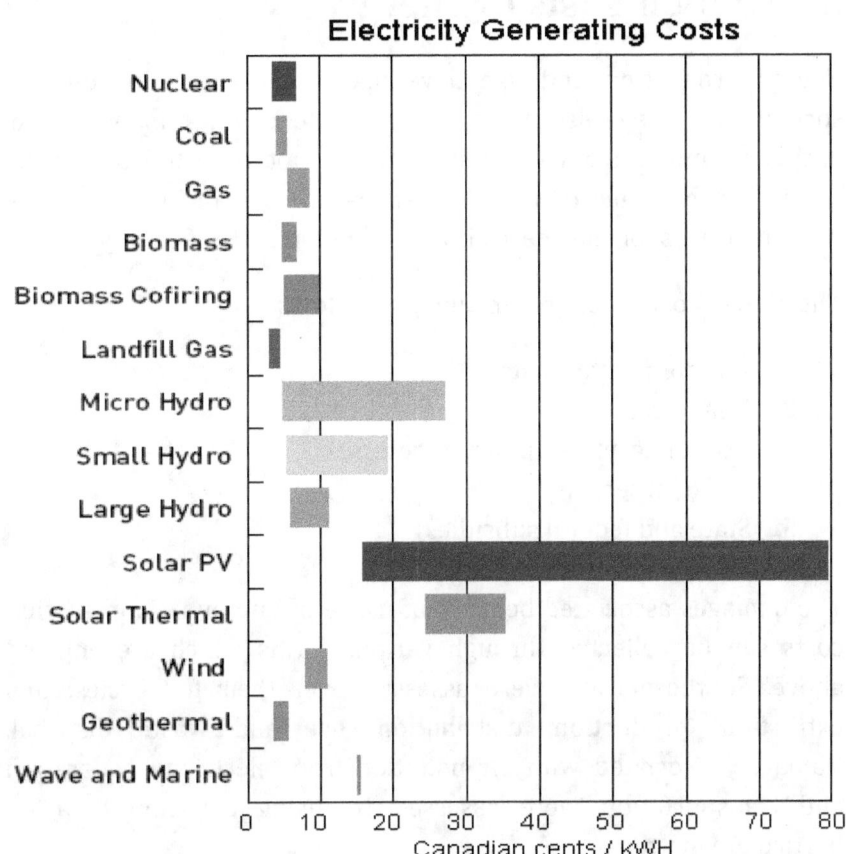

Figure 3.2 Comparison of production cost per unit for various sources.

3.2 Hidden Costs of Energy

Energy production and use have many well-known benefits to society, but they also have adverse effects on human health, property, and the environment that are not reflected in market prices which means that the prices that are not included that the consumer pays for the energy is called hidden costs of energy.

The hidden costs of energy included as follows

1. Environmental costs
2. Health costs
3. Economic and employment costs
4. National security
5. State and federal subsidies
6. Corrosion costs

The damages associated because of usage of energy resources, that costs can be collected through various forms which are enlisted above. So these costs were assessed over their full cycles; fuel extraction, production, distribution, use and waste disposal. Damages associated with air pollution from electricity generation and transportation were assessed by using emission data on particulate matter.

Sources for Hidden cost of energy: These costs ascertain about the approximate paying by the consumer to the government in the form of taxes.

1. Corrosion: Damage of the metal buildings due to corrosion were estimated to be $ 2 billion/year
2. Environmental Costs: The Environmental Protection Agency last year reported that ozone pollution was reducing crop yields by 12% ($2.5 to $3.0 billion per year). This report was recently updated, through work by the Boyce Thompson Institute for Plant Research at Cornell University. The revised estimate was that crop losses could be as high as 30% (or

$6.25 to $7.5 billion/year). The cost of controlling emissions from all coal plants to range between $4 to $6 billion per year causing utility rates, across the country, to increase between 4% and 15%.

3. Health Impacts: The major health impacts are caused by our combustion of fossil fuels. Here the information is far from complete and the maximum estimate is, in my opinion, most certainly low. This estimate is based on the combustion of low sulfur coal and the resultant sulfur dioxide pollution (which eventually winds up as acid rain). Shepard Buchanan, Bonneville Power Administration, estimates that 80% to 95% of local economic damage attributable to fossil fuels is associated with health risks. Even with low sulfur coal and remote siting, the levelized environmental costs associated with health risks would range between $.002 and $.01/kwh. The health costs associated with utilizing 20.1 quad of coal per year health costs, assuming all low sulfur and remote siting) are between $11.78 and $18.89 billion per year. Estimates from the American Lung Association indicate SO2 impact alone, could reach $82 billion. Range is between $11.8 and $82 billion.

4. Disposal of Nuclear wastes: The estimated disposal costs from nuclear wastes in a range between $1.44 billion and $8.61 billion per year.

5. Military: Our dependence on imported oil requires that our military keep the international supply lines open. The U.S. military is spending between 14.6 and 54 billion dollars yearly just defending the oil supplies coming from the Persian Gulf. On the low side, the National Defense Council places the Persian Gulf military cost at 14.6 billion. On the high side, the estimate of 54 billion is made by the Rocky Mtn. Institute. Retired Secretary of the Navy, John Lehman was quoted in Newsweek magazine as estimating the Persian Gulf military cost at 40 billion dollars yearly. And these cost

estimates only concern the Persian Gulf. There are also other hidden national security costs. One of these is military aid to oil

6. Crop losses: The EPA reported in 1988 that ozone pollution alone is reducing crop yields by up to 12% yearly, and that's about 3 billion dollars annually. Boyce Thompson of the Institute for Plant Research at Cornell University has revised this estimate to a 30% crop loss yearly with an annual price tag of 7 billion dollars. And these estimates do not include crop losses due to global warming, acid rain, and other energy related forms of air/water pollution.
Hidden Costs of Energy
Issue Hidden Cost ($bil/yr)
A. Corrosion 2.0
B. Crop Losses 2.5-7.5
C. Health Impacts 11.8-82.0
D. Radioactive Waste 4.3-31.2
E. Military 14.6-54.0
F. Subsidies 43.3-55.2
G. Employment 30.6
Total 109.2-258.0

(Source of this data: Societal costs of energy by Michael Nicklas)

3.3 Cost Analysis

Basics of Energy Estimation:

The energy costs can be estimated by having the following two information:

- The total number of watts
- The cost per kilowatt hour (Kwh) your power company charges you

Every electrical appliance or object that takes electricity to operate will be marked - on the item itself, on its packaging, or in its instruction booklet/information sheet - with either the watts it uses or with amps and volts.

To find out watts:

- Look for the watts on the item, package, or information sheet/booklet that came with it.
- If there is no watts figure, but there is amps and volts (often appearing as VAC), simply multiply the amps by the volts to get the number of watts.

On large items, like transformers, power strips, timers, etc., find the information molded (embossed) on the object or appearing on a plate or sticker on the object. The power strip for example, says 15 A 125 VAC 60 HZ.

To find the watts in this type of situation, simply multiply 15 (amps) x 125 (volts) to get 1875 watts. In the case of the strip, this is the total load it can carry. (As a couple of engineering folks pointed out to me, empty power strips consume a negligible amount.) Light bulbs are similarly easy: the watts is printed right on them. So, 100 watt light bulb uses 100 watts and 15 watt UVB-fluorescent uses 15 watts.

Formula for Estimating Energy Consumption

(Wattage × Hours Used Per Day) ÷ 1000 = Daily Kilowatt-hour (kWh) consumption

1 kilowatt (kW) = 1,000 Watts

Once you have the watts for each and every item and your power company's charge per kilowatt hour (Kwh), just run the numbers:

To find the cost to run one item:

1. Total up the watts per day for the item to get total watts/day.

2. Divide total watts/day by 1000 to get the total kilowatt hours (Kwh) per day.

3. Multiply the Kwh/day times the cost per Kwh* to get cost/day.

4. To find out the average cost per month, multiply the cost/day by 30. (Power company monthly billing cycles vary from 28-33 days)

If you want to find the cost of all items:

1. Total up all the watts per day for each item to get total watts/day.

2. Divide total watts/day by 1000 to get the total kilowatt hours (Kwh) per day.

3. Multiply the Kwh/day times the cost per Kwh to get your total cost/day.

4. To find out the average total cost per month, multiply the total cost/day by 30. (Power company monthly billing cycles vary from 28-33 days)

Examples:

Window fan:

(200 Watts × 4 hours/day × 120 days/year) ÷ 1000
= 96 kWh × 8.5 cents/kWh
= \$8.16/year

Personal Computer and Monitor:

(150 Watts × 4 hours/day × 365 days/year) ÷ 1000
= 394 kWh × 8.5 cents/kWh
= \$33.51/year

Wattage

The wattage of most appliances stamped on the bottom or back of the appliance, or on its nameplate. The wattage listed is the maximum power drawn by the appliance. Since many appliances have a range of settings (for example, the volume on a radio), the actual amount of power consumed depends on the setting used at any one time.

If the wattage is not listed on the appliance, you can still estimate it by finding the current draw (in amperes) and multiplying that by the voltage used by the appliance. Most appliances in the United States use 120 volts. Larger appliances, such as clothes dryers and electric cooktops, use 240 volts. The amperes might be stamped on the unit in place of the wattage. If not, find a clamp-on ammeter an electrician's tool that clamps around one of the two wires on the appliance to measure the current flowing through it. You can obtain this type of ammeter in stores that sell electrical and electronic equipment. Take a reading while the device is running; this is the actual amount of current being used at that instant. When measuring the current drawn by a motor, note that the meter will show about three times more current in the first second that the motor starts than when it is running smoothly.

Many appliances continue to draw a small amount of power when they are switched "off." These "phantom loads" occur in most appliances that use electricity, such as VCRs, televisions, stereos, computers, and kitchen appliances. Most phantom loads will increase the appliance's energy consumption a few watt-hours. These loads can be avoided by unplugging the appliance or using a power strip and using the switch on the power strip to cut all power to the appliance.

Typical Wattages of Various Appliances

Here are some examples of the range of nameplate wattages for various household appliances:

- Aquarium = 50–1210 Watts
- Clock radio = 10
- Coffee maker = 900–1200
- Clothes washer = 350–500
- Clothes dryer = 1800–5000
- Dishwasher = 1200–2400 (using the drying feature greatly increases energy consumption)
- Dehumidifier = 785
- Electric blanket- *Single/Double* = 60 / 100
- Fans
  - Ceiling = 65–175
  - Window = 55–250
  - Furnace = 750
  - Whole house = 240–750
- Hair dryer = 1200–1875

- Heater *(portable)* = 750–1500
- Clothes iron = 1000–1800
- Microwave oven = 750–1100
- Personal computer
  - CPU - awake / asleep = 120 / 30 or less
  - Monitor - awake / asleep = 150 / 30 or less
  - Laptop = 50
- Radio *(stereo)* = 70–400
- Refrigerator *(frost-free, 16 cubic feet)* = 725
- Televisions (color)
  - 19" = 65–110
  - 27" = 113
  - 36" = 133
  - 53"-61" Projection = 170
  - Flat screen = 120
- Toaster = 800–1400
- Toaster oven = 1225
- VCR/DVD = 17–21 / 20–25
- Vacuum cleaner = 1000–1440
- Water heater *(40 gallon)* = 4500–5500
- Water pump *(deep well)* = 250–1100

Review Questions

1. Explain in detail the economical aspects of various methods of power generation?
2. Explain economical aspects of electrical energy generation. Differentiate costs of electricity generation from various energy sources.
3. Name few hidden cost of energy.
4. Define Hidden cost of energy?
5. What are the hidden costs involved in energy production and how do you quantify them?
6. What do you require to estimate the cost of energy consumption? And estimate the total energy consumption and the cost of total energy consumption of your home?
7. Calculate electricity consumption and the cost in your home?
8. Consider your home containing 5 ceiling fans, 3 Incandescent lamps, 6 CFL bulbs, 1 refrigerator, 1 geyser, 1 television, 1 desktop computer, 6 tube lights, 2 bed lamps,1 mixer,2 exhaust fans, and 1 laptop. Estimate the cost of electricity per month assuming necessary data suitably.
9. Compare different forms of energy generation in economical point of view?
10. Give the details of cost of unit price of all the resources.
11. Why can't we change to solar energy production as a major commercial source instead of fossil fuel based, explain in terms of economics?
12. Write a short note on Hidden cost of energy and its impact on society.
13. Analyze the usage of energy in your home and give your ideas of energy conservation.

Objective type questions

1. Which among the followings having less cost of electricity power generation?

a) Solar photo voltaic   b) Solar thermal   c) wind   d) Nuclear

2. The world faces an energy crisis because _____

    a) World demand for energy will increase
    b) World oil production will peak and begin to decline
    c) Shortages and the resulting escalation of prices can shock the economic and political order
    d) All of the above

3. Running cost of a hydro plant with respect to that of a thermal plant is

    a)   More
    b)   Less
    c)   Varying from installation to installation
    d)   Depends on location

4. Hidden costs means-------

    a)   Taxes
    b)   Prices not included in commodities
    c)   Prices paid by individual interests
    d)   Both a and b

This chapter deals with the harmful gases released while the energy production either from conventional or unconventional resources. These gases presence in the atmosphere will adversely affects the environment and also leads to change in the climatic conditions. This chapter was divided into four sections viz., Environmental problems, Impact of energy on environment, Carbon footprint and Ecological footprint. In the first section, clearly bringing out the environmental problems like global warming (i.e., CO_2 intensity in the atmosphere), acid rains and ozone depletion due to production of electricity. In the second section dealing with the emission of malicious gases from renewable and non renewable resources. In the third section discussing about the emission of CO_2 directly or indirectly into the atmosphere by human made activities and calculating per capita of carbon dioxide emission by usage of electricity. Finally in the fourth section deals with the land required to survive man's life to overcome these environmental problems.

4.1 Environmental Aspects of Energy

The environmental issues concerning with combustion of fossil fuels are more in present panorama. Excessive fossil fuel energy use not only has caused severe and growing damage to the environment from greenhouse gas emissions and oil spills, but also has brought political crises to countries in the form of global resource conflicts and food shortages. Energy use and distribution is of basic importance to social issue, with the possible exception of agriculture and forestry, has made the greatest impact on the environment by human activity as a result of the large scale and wide spread nature of energy related activities. Although the terms energy and environment are of prominent role, in usage of energy resources as well as to protect the environment. But, because of energy that leads to some environmental problems with emission of toxic gases while extraction, production and transportation and they have now widened to cover regional and global issues such as acid rain and the greenhouse effect. Such problems have now become major political issues and the subject of international debate and regulation.

Global Environmental problems:

One of the most important aspects of global issues is that it affects all mankind on a global scale without regard to any particular to country or region. Some of the key environmental problems are

- ✓ Global warming
- ✓ Acid Rain
- ✓ Ozone Depletion

Global Warming:

The Earth is warming because of manmade activities. When CO_2 and other heat-trapping emissions are released into the air, these gases holding heat in the atmosphere and warming the planet. Overloading our atmosphere with carbon has far-reaching effects for people all around the world recording high temperatures, more extreme storms, more severe droughts, deadly heat waves, rising sea levels, and more acidic oceans which can affect the basic needs like food and water supply, endanger our health, menace our national security, and threaten other basic human requirements.

However, there is much we can do to protect the health and economic well-being of current and future generations from the consequences of the heat-trapping emissions caused when we burn coal, oil, and gas to generate electricity and also for transportation purposes. Tropical deforestation is another major contributor of global warming. When these forests are burned, they release huge amounts of carbon into the atmosphere. Because of extinction of forests there will be no longer available to absorb CO_2. Here is the urgency to significantly reduce the amount of heat-trapping emissions. As individuals, we can help by being mindful of our electricity use, driving more efficient cars, preferring going by mass transportation, reducing deforestation, reducing the number of miles we drive, and taking other steps to reduce our own consumption of fossil fuels.

To limit sea level rise to only 1 m and species loss to 20% by the end of this century, additional warming must be limited to 1°C. This means stabilizing atmospheric CO_2 at about 450–500 ppm. The United States is the second largest emitter of CO_2 emissions after China. The United States currently emits 23% of global CO_2 and needs to reduce CO_2 by 60 to 80% by midcentury.

Acid Rain:

Acid deposition can occur via natural sources like volcanoes but it is mainly caused by the release of sulfur dioxide and nitrogen oxide during fossil fuel combustion. When these gases are discharged into the atmosphere they react with the water, oxygen, and other gases already present there to form sulfuric acid, ammonium nitrate, and nitric acid. These acids then disperse over large areas because of wind patterns and fall back to the ground as acid rain or other forms of precipitation. The gases responsible for acid deposition are normally a byproduct of electric power generation and the burning of coal.

There are several important impacts of acid deposition on both natural and man-made environments. Aquatic settings are the most clearly impacted by acid deposition though because acidic precipitation falls directly into them. Both dry and wet deposition also runs off of forests, fields, and roads and flows into lakes, rivers, and streams. As this acidic liquid flows into larger bodies of water, it is diluted but over time, acids can accrue and lower the overall pH of the body. Acid deposition also causes clay soils to release aluminum and magnesium further lowering the pH in some areas. If the pH of a lake drops below 4.8, its plants and animals risk death. Aside from aquatic bodies, acid deposition can significantly impact forests. As acid rain falls on trees, it can make them lose their leaves, damage their bark, and stunt their growth. By damaging these parts of the tree, it makes them vulnerable to disease, extreme weather, and insects. Acid falling on a forest's soil is also harmful because it disrupts soil nutrients, kills microorganisms in the soil, and can sometimes cause a calcium deficiency. Trees at high altitudes are also susceptible to problems induced by acidic cloud cover as the moisture in the clouds blankets them. Damage to forests by acid rain is seen all over the world, but the most advanced cases are in Eastern Europe. It's estimated that in Germany and Poland, half of the forests are damaged, while 30% in Switzerland have been affected.

Finally, acid deposition also has an impact on architecture and art because of its ability to corrode certain materials. As acid lands on buildings (especially those constructed with limestone) it reacts with minerals in the stones sometimes causing it to disintegrate and wash away. Acid deposition can also corrode modern buildings, cars, railroad tracks, airplanes, steel bridges, and pipes above and below ground. Because of these problems and the adverse effects air pollution has on human health, a number of steps are being taken to reduce sulfur and nitrogen emissions. Most notably, many governments are now requiring energy producers to clean smoke stacks by using scrubbers which trap pollutants before they are released into the atmosphere and catalytic converters in cars to reduce their emissions. Additionally, alternative energy sources are gaining more prominence today and funding is being given to the restoration of ecosystems damaged by acid rain worldwide.

Ozone layer depletion:

Ozone is present in the stratosphere that it protects the Earth from the ultraviolet rays passed by the sun. The stratosphere is 30 miles above the earth and at the very top it contains ozone. If the ozone layer is depleted by human action, the effects on the planet could be catastrophic. Ozone is a bluish gas that is formed by three atoms of oxygen. The form of oxygen that humans breathe in consists of two oxygen atoms, O_2. When found on the surface of the planet, ozone is considered a dangerous pollutant and is one substance responsible for producing the greenhouse effect.

The highest regions of the stratosphere contain about 90% of all ozone. In recent years, the ozone layer has been the subject of much discussion. And rightly so, because the ozone layer protects both plant and animal life on the planet. The fact that the ozone layer was being depleted was discovered in the mid-1980s. The main cause of this is the release of CFCs, chlorofluorocarbons. There, the chlorine atom is removed from the CFC and attracts one of the three oxygen atoms in the ozone molecule. The process continues, and a single chlorine atom can destroy over 100,000 molecules of ozone. Antarctica was an early victim of ozone destruction. A massive hole in the ozone layer right above Antarctica now threatens not only that continent, but many others that could be the victims of Antarctica's melting icecaps. In the future, the ozone problem will have to be solved so that the protective layer can be conserved. Because of exposing to UV rays causes skin cancer. In addition to cancer, some research shows that a decreased ozone layer will increase rates of malaria and other infectious diseases. The environment will also be negatively affected by ozone depletion. The life cycles of plants will change, disrupting the food chain. Effects on animals will also be severe, and are very difficult to foresee.

Oceans will be hit hard as well. The most basic microscopic organisms such as plankton may not be able to survive. If that happened, it would mean that all of the other animals that are above plankton in the food chain would also die. Other ecosystems such as forests and deserts will also be harmed. The planet's climate could also be affected by depletion of the ozone layer. Wind patterns could change, resulting in climatic changes throughout the world.

The discovery of the ozone depletion problem came as a great surprise. Now, action must be taken to ensure that the ozone layer is not destroyed. Because CFCs are so widespread and used in such a great variety of products, limiting their use is hard. Also, since many products already contain components that use CFCs, it would be difficult if not impossible to eliminate those CFCs already in existence. The CFC problem may be hard to solve because there are already great quantities of CFCs in the environment. CFCs would remain in the stratosphere for another 100 years even if none were ever produced again.

Despite the difficulties, international action has been taken to limit CFCs. In the Montreal Protocol, 30 nations worldwide agreed to reduce usage of CFCs and encouraged other countries to do so as well.

4.2   Impact of Energy on Environment

Coal:

Coal mining has the potential to harm air, water and land quality if it is not done with proper care. Acidic water may drain from abandoned mines underground, and the burning of coal causes the emission of harmful materials including carbon dioxide, sulfur dioxide and mercury. "Clean coal" technology is being developed to remove harmful materials before they can affect the environment, and to make it more energy-efficient so less coal is burned. The coal industry also restores mined land to or prepares it for more productive uses once surface mining is done.

Petroleum (Oil and Gas):

Great strides have been made to ensure that oil and gas producers make as little impact as possible on the natural environments in which they operate. These include drilling multiple wells from a single location to minimize damages to the surface, using environmentally sound chemicals to stimulate well production and restoring the surface as nearly as possible to pre-drilling conditions (as required by landowners and state or federal agencies, who often must approve the company's completion of restoration activities).

When many people think of oil and the environment, they think of oil spills. The reality is that the exploration and production of oil rarely create an oil spill. For decades the offshore oil and gas industry has had a strong safety and environmental record in operating in the Gulf of Mexico, with less than 0.001 percent of the oil produced in federal waters spilled since 1980. The Deepwater Horizon event is a stark reminder of the risks and challenges in offshore operations. The oil and gas industry takes safety and environmental stewardship very seriously. That is why when it is known how the Deepwater Horizon accident occurred, the industry will work together to ensure that that this kind of event never happens again anywhere in the world.

Most oil spills occur primarily during transportation, mostly involving the tankers that are used to move oil from where it is produced to where consumers need it. But oil spills from transportation have declined significantly during the past few years, and the growing use of double-hulled tankers provides extra protection. Another source of oil spills during transportation is pipelines. Unfortunately, a major reason for spills from pipelines in developing countries is civil unrest. Weather, such as hurricanes, is another factor in pipeline-related spills.

Urban runoff and natural seeps are large sources of oil pollution. Urban runoff comes from rain washing away oil drips from cars or machinery and people pouring used oil into the gutter and using other improper disposal methods. Natural seepage is actually the largest single source of petroleum inputs in marine environments totaling 47%. When burned, petroleum products emit carbon dioxide, carbon monoxide and other air toxins, all of which have a negative effect on the environment.

Biofuels: Biomass, Ethanol and Biodiesel:

On the surface, biofuels look like an ideal energy solution. Since plants absorb carbon dioxide as they grow, crops could counteract the carbon dioxide released by cars. They are also renewable, and can be planted to replenish supplies. Unfortunately, it's not that easy.

It takes a tremendous amount of energy to grow crops, make fertilizers and pesticides and process plants into fuel. There is ongoing debate if ethanol from corn provides more energy than it uses for growing and processing the plants. Also, fossil fuels provide much of the energy in biofuels production, so biofuels may not replace as much oil as they use.

Biomass creates harmful emissions like carbon dioxide and sulfur when it is burned, but causes less pollution than fossil fuels. Even burning wood in a fireplace or stove can create pollutants like carbon monoxide. Burning municipal solid waste, or garbage that would otherwise go into a landfill, can also cause potentially dangerous emissions. Combustion of these materials must be carefully controlled. Disposing of the resulting ash can also pose a problem, as it may contain harmful metals like lead and cadmium.

Ethanol is often added to gasoline, and while these mixtures burn cleaner than pure gasoline, they also have higher "evaporative emissions" from dispensing equipment and fuel tanks. These emissions contribute to ozone problems and smog. Burning ethanol also creates carbon dioxide.

Biodiesel creates less sulfur oxides, particulate matter, carbon monoxide and hydrocarbons when burned that traditional petroleum diesel. But biodiesel creates more nitrogen oxide than petroleum diesel.

Geothermal:

Geothermal power plants have relatively little environmental impact and they burn no fuel to create electricity. These plants do create small amounts of carbon dioxide and sulfur compounds, but geothermal emissions are far smaller than those created by fossil fuel power plants.

Hydropower:

While hydropower does not cause water or air pollution, it does have an environmental impact: Hydroelectric power plants may harm fish populations, change water temperature and flow (disturbing plants and animals) and force the relocation of people and animals who live near the dam site. Some fish, like salmon, may be prevented from swimming upstream to spawn. Technologies like fish ladders help salmon go up over dams and enter upstream spawning areas, but the presence of hydroelectric dams changes their migration patterns and hurts fish populations. Hydropower plants can also cause low dissolved oxygen levels in the water, which is harmful to river habitats. Reservoirs may also lead to the creation of methane, a harmful greenhouse gas.

Solar Energy:

Solar energy produces no air or water pollution or greenhouse gases. However, it has some indirect impacts on the environment. For example, the manufacturing of photovoltaic cells (PV) produces some toxic materials and chemicals. Ecosystems can also be affected by solar systems. Water from underground wells may be required to clean concentrators and receivers, and to cool the generator, which may harm the ecosystem in dry climates.

Uranium (Nuclear Energy):

Nuclear power plants produce no air pollution or carbon dioxide, but they do produce byproducts like nuclear waste and spent fuels. Most nuclear waste is low level (for example, disposable items that have come into contact with small amounts of radioactive dust), and special regulations are in place to prevent them from harming the environment. But some spent fuel is highly radioactive and must be stored in specially designed facilities. In addition to the fuel waste, much of the equipment in the nuclear power plants becomes contaminated with radiation and will become radioactive waste after the plant is closed. These wastes will remain radioactive for many thousands of years, which may not allow re-use of the contaminated land.

Nuclear power plants use large quantities of water for steam production and for cooling, affecting fish and other aquatic life. Likewise, heavy metals and salts can build up in the water used in the nuclear power plant systems. When water is discharged from the power plant, these pollutants can negatively affect water quality and aquatic life.

Wind Energy:

Wind is a clean energy source. It produces no air or water pollution because no fuel is burned to generate electricity. The most serious environmental impact from wind energy may be its effect on bird and bat mortality. Wind turbine design has changed dramatically in the last couple of decades to reduce this impact. Turbine blades are now solid, so there are no lattice structures that entice birds to perch. Also, the blades' surface area is much larger, so they don't have to spin as fast to generate power. Slower-moving blades mean fewer bird collisions.

4.3   Climate Changes

Climate change, first and foremost, is a consequence of the high use of fossil fuel. Even though climate change is a global problem, the fossil fuel dependence that contributes to it carries growing economic risks for the emitting country. Working our way out of this addiction takes time, and the longer we wait to radically rethink and retool our societies, the less chance we will have to alter course.

World leaders have, to a great extent, affirmed the need to stay within a 2 degrees Celsius climate alteration (at a minimum) to avoid widespread calamity. This means reducing carbon concentrations in the atmosphere to between 450 (by optimistic estimates) and 350 parts per million (by more realistic estimates). Reaching even the more relaxed target will require a massive shift away from fossil fuel now (and not in a decade or two) and a wholesale restructuring of the way we produce and use energy. Yet hardly anybody admits this mathematical truth.

Effects of Climate Change Today

Over 100 years ago, people worldwide began burning more coal and oil for homes, factories, and transportation. Burning these fossil fuels releases carbon dioxide and other greenhouse gases into the atmosphere. These added greenhouses gases have caused Earth to warm more quickly than it has in the past.

How much warming has happened? Scientists from around the world with the Intergovernmental Panel on Climate Change (IPCC)

---

tell us that during the past 100 years, the world's surface air temperature increased an average of 0.6° Celsius (1.1°F). This may not sound like very much change, but even one degree can affect the Earth. Below are some effects of climate change that we see happening now.

- ✓ **Arctic sea ice is melting:** The summer thickness of sea ice is about half of what it was in 1950. Melting ice may lead to changes in ocean circulation. Plus melting sea ice is speeding up warming in the Arctic.
- ✓ **Glaciers and permafrost are melting.** Over the past 100 years, mountain glaciers in all areas of the world have decreased in size and so has the amount of permafrost in the Arctic. Greenland's ice sheet is melting faster too.
- ✓ **Sea-surface temperatures are warming.** Warmer waters in the shallow oceans have contributed to the death of about a quarter of the world's coral reefs in the last few decades. Many of the coral animals died after weakened by bleaching, a process tied to warmed waters.
- ✓ **The temperatures of large lakes are warming.** The temperatures of large lakes world-wide have risen dramatically. Temperature rises have increased algal blooms in lakes, favor invasive species, increase stratification in lakes and lower lake levels.
- ✓ **Heavier rainfall cause flooding in many regions.** Warmer temperatures have led to more intense rainfall events in some areas. This can cause flooding.
- ✓ **Extreme drought is increasing.** Higher temperatures cause a higher rate of evaporation and more drought in some areas of the world.
- ✓ **Crops are withering.** Increased temperatures and extreme drought are causing a decline in crop productivity around the world. Decreased crop productivity can mean food shortages which have many social implications.

- ✓ **Ecosystems are changing**. As temperatures warm, species may either move to a cooler habitat or die. Species that are particularly vulnerable include endangered species, coral reefs, and polar animals. Warming has also caused changes in the timing of spring events and the length of the growing season.
- ✓ **Hurricanes have changed in frequency and strength.** There is evidence that the number of intense hurricanes has increased in the Atlantic since 1970. Scientists continue to study whether climate is the cause.
- ✓ **More frequent heat waves.** It is likely that heat waves have become more common in more areas of the world.
- ✓ **Warmer temperatures affect human health.** There have been more deaths due to heat waves and more allergy attacks as the pollen season grows longer. There have also been some changes in the ranges of animals that carry disease like mosquitoes.
- ✓ **Seawater is becoming more acidic.** Carbon dioxide dissolving into the oceans, is making seawater more acidic. There could be impacts on coral reefs and other marine life.

Health impacts of climate change

- ✓ Climate change is a major problem caused by the increase of human activities leading to several direct and indirect impacts on health. The combustion of fossil fuels, increasing number of industries, and large-scale deforestation are some of the causes for the accumulation of GHGs (greenhouse gases) in the atmosphere. According to the IPCC (Intergovernmental Panel on Climate Change), an increase in carbon dioxide and other GHGs, like methane, ozone, nitrous oxide, and chlorofluorocarbons, in the atmosphere is expected to increase the average global temperature by 1.5 ° C to 4.5 ° C. This in turn will lead to changes in rainfall and snowfall, more intense or frequent droughts, floods, and storms, as well as a rise in sea level. These climatic changes will have wide-ranging harmful effects including increase in heat-related mortality, dehydration, spread of infectious diseases, malnutrition, and damage to public health infrastructure. Thus we should take appropriate measures to stop this climate change.

✓ The weather has a direct impact on our health. If the overall climate becomes warmer, there will be an increase in health problems. It is anticipated that there will be an increase in the number of deaths due to greater frequency and severity of heat waves and other extreme weather events. The elderly, the very young and those suffering from respiratory and cardiovascular disorders will probably be affected by such weather extremes as they have lesser coping capacity. An extreme rise in the temperature will affect people living in the urban areas more than those in the rural areas. This is due to the 'heat islands' that develop here owing to the presence of concrete constructions, paved and tarred roads. Higher temperatures in the cities would lead to an increase in the ground-level concentration of ozone thereby increasing air pollution problems.

✓ Indirectly, changes in weather pattern, can lead to ecological disturbances, changes in food production levels, increase in the distribution of malaria, and other vector-borne diseases. Fluctuation in the climate especially in the temperature, precipitation, and humidity can influence biological organisms and the processes linked to the spread of infectious diseases.

✓ Higher temperature will cause the sea levels to rise that could lead to erosion and damage to important ecosystems such as wetlands and coral reefs. Direct impact of this rise would include deaths and injury caused by intense flooding. Temperature rise would indirectly result in geohydrological changes along the coastline such as saltwater intrusion into the groundwater and the wetlands, coral reef destruction, and damage to the drainage in the low-lying areas. Climate change could increase air pollution levels by accelerating the atmospheric chemical reactions that produce photochemical oxidants due to a rise in the temperature.

✓ Due to global warming there will be an increase in the areas of habitat of disease-spreading insects such as the mosquito, causing an increase in the transmission of infection through these carriers.

✓ Potential effects on health due to sea level rise include: death and injury due to flooding; reduced availability of fresh water due to saltwater intrusion; contamination of water supply through pollutants from submerged waste dumps; change in the distribution of disease-spreading insects; effect on the nutrition due to a loss in agriculture land and changes in fish catch; and

4.4 Carbon Footprint

A **carbon footprint** is a measure of the impact our activities have on the environment, and in particular climate change. It relates to the amount of greenhouse gases produced in our day-to-day lives through burning fossil fuels for electricity, heating and transportation. The carbon footprint is a measurement of all greenhouse gases we individually produce and has units of tons (or kg) of carbon dioxide equivalent.

Each of our everyday actions consumes energy and produces carbon dioxide emissions e.g. taking flights, driving our cars, heating or cooling offices. Carbon Offsets can be used to compensate for the emissions produced by funding an equivalent carbon dioxide saving elsewhere. In other words: When you drive a car, the engine burns fuel which creates a certain amount of CO_2, depending on its fuel consumption and the driving distance. (CO_2 is the chemical symbol for carbon dioxide). When you heat your house with oil, gas or coal, then you also generate CO_2. Even if you heat your house with electricity, the generation of the electrical power may also have emitted a certain amount of CO_2. When you buy food and goods, the production of the food and goods also emitted some quantities of CO_2.

The carbon footprint of an individual person is the summation of carbon dioxide gas emitted based on your daily activities in a given time frame. Usually a carbon footprint is calculated for the time period of a year. Few people express their carbon footprint in kg carbon rather than kg carbon dioxide. You can always convert kg carbon dioxide in kg carbon by multiplying with a factor 0.27 (1000 kg CO_2 equals 270 kg carbon). The carbon footprint is a very powerful tool to understand the impact of personal behavior on global warming. Most people are shocked when they see the amount of CO_2 their activities create! If you personally want to contribute to stop global warming, the calculation and constant monitoring of your personal carbon footprint is essential. Even there are carbon calculators to calculate your carbon footprint by giving your daily usage of various energy resources like electricity, transportation, hospitality, consumption of food etc., for instance the more processed food, clothes and furniture means you are contributing to the more carbon footprint indirectly.

The carbon footprint comes in two ways.

1. Primary foot print - which monitors your carbon emission directly (through energy production while combustion of fossil fuels).
2. Secondary foot print relates to your indirect carbon emissions (such as through your food preference, fashion, recreation, purchasing of any goods etc).

Food Preference - Meat lovers emit more carbon than vegetarians as the process of preparing the meat is more than producing organic vegetables. Fashion - If you are buying clothes and accessories to keep us with the fashion trends etc. again you are emitting more carbon. Recreation - walking, cycling and swimming emit less or no carbon compared to Speed Boating. Therefore for you can start with recycling which is the most common form. Reuse is also a good method, then start going green by reducing the plastic bottles and polythene then try to reduce the amount of processed food you eat and go organic and even planting trees to reduce the carbon foot print is a great initiative.

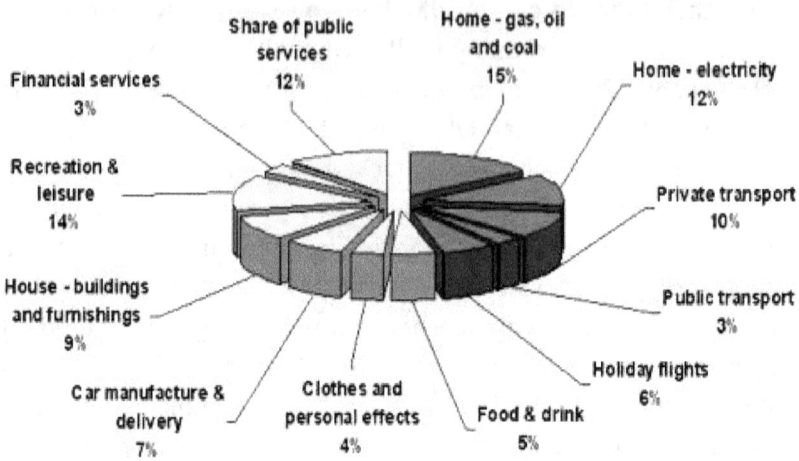

Figure 4.1: The pie chart shows the main elements of which made by person's carbon footprint in developed countries.

(With courtesy the above data taken from Carbon Footprint Ltd serve businesses and organization website)

Each of the following activities add 1 kg of CO_2 to your personal carbon footprint:

- Travel by public transportation (train or bus) a distance of 10 to 12 km (6.5 to 7 miles)
- Drive with car a distance of 6 km or 3.75 miles(assume 7.3 litres petrol per 100 km or 39 mpg)
- Fly with a plane a distance of 2.2 km or 1.375 miles.
- Operate your computer for 32 hours (60 Watt consumption assumed)
- Production of 5 plastic bags
- Production of 2 plastic bottles
- Production of 1/3 of an American cheeseburger (yes, the production of each cheeseburger emits 3.1 kg of CO_2)

Calculation of carbon footprint:

The volume of air breathed in and out by an adult male walking or jogging at 4 miles per hour is about 35 litres per minute, compared with less than 10 litres per minute while at rest.

(Source - California Environmental Protection Agency. Research Note 94-11)

1. Residential:

Electricity: On average, electricity sources emit 1.306lbs CO_2 per kWh *(Source: EPA's eGrid)*

Heating Oil: There are 22.377 pounds of CO_2 per gallon of heating oil (diesel fuel). We multiply 22.377 by the number of gallons of heating oil consumed annually and divide by 2,204.6 to get metric tons of CO_2 per gallon. *(Source: US DOE 1605(b) Voluntary Reporting of Greenhouse Gases Program)*

2. Transportation:

Vehicles: Unleaded gasoline has 8.91 kg (19.643lbs) of CO_2 per gallon. By dividing number of miles driven by miles per gallon, we get the number of gallons of gasoline consumed annually. We multiply this by 8.91 and divide by 1,000 to get metric tons of CO_2. *(Source: US DOE 1605(b) Voluntary Reporting of Greenhouse Gases Program)*

Air Travel: CO_2 emissions in air travel vary by length of flight-- ranging from 0.60kg CO_2 per passenger mile to 0.40kg CO_2 per passenger mile depending on the flight distance. Our calculator allows the user to take the issue of radiative forcing into account. *(Sources: EPA Climate Leaders)*

Rail Travel: The CO_2 emissions for rail travel vary by distance of the trip. On average, commuter rail and subway trains emit 0.35 lbs CO_2 per passenger mile, and long distance trains emit 0.42 lbs

CO_2 per passenger mile *(Source WRI: Employee commuting spreadsheet)*

Bus Travel: The CO_2 emission associated with bus travel vary by distance of trip. Inner city commuting buses emit 0.66 lbs CO_2 per passenger mile, and long distance bus trips emit 0.18 lbs CO_2 per passenger mile *(Source WRI: Employee commuting spreadsheet)*

3. Hospitality:

Meals: The average person's diet contributes 2,920 kilograms CO_2e to the atmosphere each year. By dividing by 365, it is deduced that a person emits on average 8 kg CO_2 per day from their meals. This calculation is based on an average US, non-vegetarian diet. The emissions for food preparation are not included in this calculation. *(Source: Johns Hopkins Bloomberg School of Public Health)*

Hotel Rooms: Emissions associated with a one night stay in a hotel room are calculated at 29.53 kg CO_2 per room day for an average hotel. For upscale hotels, emissions are calculated at 33.38 kg CO_2 per room day *(Source: Environmental Protection Agency)*. CHP Potential in the Hotel and Casino Market Sectors, prepared by Energy and Environmental Analysis, Inc. for EPA.

Reduction of Carbon Footprint:

For Individuals

Here's a list of simple of things that individuals has to practice in order to reduce their own carbon footprint.

- ✓ Turn off appliances when not in use (lights, television, DVD player, computer etc)

- ✓ Turn down the central heating slightly (try just 1 to 2 degrees C). Just 1 degree will help reduce your heating bill by about 8%.

- ✓ Turn down the water heating setting (just 2 degrees will make a significant saving)
- ✓ Check the central heating timer setting - remember there is no point heating the house after you have left for work
- ✓ Fill your dish washer and washing machine with a full load - this will save you water, electricity, and washing powder
- ✓ Fill the kettle with only as much water as you need
- ✓ Do your weekly shopping in a single trip
- ✓ Hang out the washing to dry rather than tumble drying it

The following is a list of items that may take an initial investment, but by reducing the energy bills that will indirectly reduce the carbon footprint.

- ✓ Fit energy saving light bulbs
- ✓ Install thermostatic valves on your radiators
- ✓ Insulate your hot water tank, your loft and your walls
- ✓ Installing cavity wall installation
- ✓ By installing 180mm thick loft insulation
- ✓ Recycle your grey water
- ✓ Replace your old fridge / freezer (if it is over 15 years old), with a new one with energy efficiency rating of "A"
- ✓ Replace your old boiler with a new energy efficient condensing boiler

Travel less and travel more carbon footprint friendly. (Source of collecting this data: Carbon footprint website)

- ✓ Car share to work, or for the kids school run
- ✓ Use the bus or a train rather than your car

- ✓ For short journeys either walk or cycle

- ✓ Try to reduce the number of flights you take

- ✓ See if your employer will allow you to work from home one day a week

- ✓ Next time you replace your car - check out diesel engines. With one of these you can even make your own Biodiesel fuel. Find out more about Biodiesel.

- ✓ When staying in a hotel - turn the lights and air-conditioning off when you leave your hotel room, and ask for your room towels to be washed every other day, rather than every day

As well as your primary carbon footprint, there is also a secondary footprint that you cause through your buying habits.

- ✓ Don't buy bottled water if your tap water is safe to drink

- ✓ Buy local fruit and vegetables, or even try growing your own

- ✓ Buy foods that are in season locally

- ✓ Don't buy fresh fruit and vegetables which are out of season, they may have been flown in

- ✓ Reduce your consumption of meat

- ✓ Try to only buy products made close to home (look out and avoid items that are made in the distant lands)

- ✓ Buy organic produce

- ✓ Don't buy over packaged products

- ✓ Recycle as much as possible

4.5 Ecological Footprint

The Ecological Footprint has emerged as the world's premier measure of humanity's demand on nature. It measures how much land and water area a human population requires to produce the resource it consumes and to absorb its carbon dioxide emissions, using prevailing technology.

The ecological footprint is a measure of human demand on the Earth's ecosystems. The Ecological Footprint uses yields of primary products (from cropland, forest, grazing land and fisheries) to calculate the area necessary to support a given activity. Biocapacity is measured by calculating the amount of biologically productive land and sea area available to provide the resources a population consumes and to absorb its wastes. Ecological footprint analysis compares human demand on nature with the biosphere's ability to regenerate resources and provide services. Footprint values at the end of a survey are categorized for carbon, food, housing, and goods and services as well as the total footprint number of earths needed to sustain the world's population at that level of consumption.

Today humanity uses the equivalent of 1.5 planets to provide the resources we use and absorb our waste. This means it now takes the Earth one year and six months to regenerate what we use in a year. Moderate UN scenarios suggest that if current population and consumption trends continue, by the 2030s, we will need the equivalent of two Earths to support us. And of course, we only have one. Turning resources into waste faster than waste can be turned back into resources puts us in global ecological overshoot, depleting the very resources on which human life and biodiversity depend.

A nation's consumption is calculated by adding imports to and subtracting exports from its national production. Results from this analysis shed light on a country's ecological impact. For example, the

National Footprint Accounts identify whether or not a country's Ecological Footprint exceeds its biocapacity. A country has an ecological reserve if its Footprint is smaller than its biocapacity; otherwise it is operating with an ecological deficit. The former are often referred to as ecological creditors, and the latter ecological debtors. Today, most countries, and the world as a whole, are running ecological deficits. The world's ecological deficit is referred to as global ecological overshoot. Overshoot, which in this context is shorthand for ecological overshoot, occurs when a population's demand on an ecosystem exceeds the capacity of that ecosystem to regenerate the resources it consumes and to absorb its carbon dioxide emissions.

Our natural systems can only generate a finite amount of raw materials (fish, trees, crops, etc.) and absorb a finite amount of waste (such as carbon dioxide emissions). Global Footprint Network quantifies this rate of output by measuring biocapacity – nature's ability to renew resources and provide ecological services. Biocapacity is as measurable as GDP – and, ultimately, far more significant, as access to basic living resources underlies every economic activity a society can undertake.

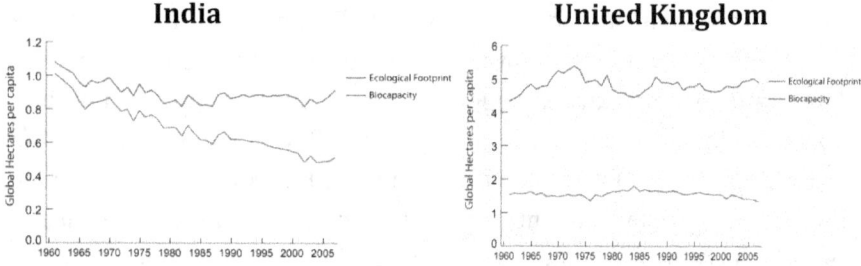

Figure 4.2: Tracks the per-person resource demand (Ecological Footprint) and resource supply (Biocapacity) in India and United Kingdom since 1961. Biocapacity varies each year with ecosystem management, agricultural practices (such as fertilizer use and irrigation), ecosystem degradation, and weather.

United States   **China**

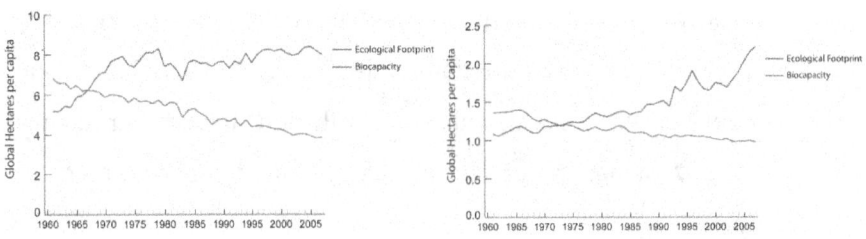

Figure 4.3: Tracks the per-person resource demand (Ecological Footprint) and resource supply (Biocapacity) in United States and China since 1961. Biocapacity varies each year with ecosystem management, agricultural practices (such as fertilizer use and irrigation), ecosystem degradation, and weather.

(Source: The above figures taken from global footprint network /Ecological footprint network)

In 2006, the average biologically productive area per person worldwide was approximately 1.8 global hectares (gha) per capita. The U.S. footprint per capita was 9.0 gha, and that of Switzerland was 5.6 gha per person, while China's was 1.8 gha per person. (Source: Living Planet Report 2008 outlines scenarios for humanity's future. Global Footprint Network. Retrieved: 2009-02-15). Ecological footprint analysis is now widely used around the globe as an indicator of environmental sustainability. It can be used to measure and manage the use of resources throughout the economy and also to explore the sustainability of individual lifestyles, goods and services, organizations, industry sectors, neighborhoods, cities, regions and nations. Since 2006, a first set of ecological footprint standards exist that detail both communication and calculation procedures.

Review Questions

1. The electricity utilization in India for domestic, agriculture and industrial purposes were mainly generated from coal fired thermal power plants which cause pollution. But why can't we go for clean technologies like wind, Hydel and solar energies as it doesn't cause any pollution. Justify your answer clearly bringing out the advantages and disadvantages of them?

2. What is climate change? How human activities related to climate change? What are the effects of climate change?

3. What are the effects of carbon dioxide levels in the atmosphere?

4. For transportation purposes we mainly depends on petrol, diesel and CNG gas which cause pollution by combustion of these fuels and also the supply and demand of these fuels are unequal. In order to overcome this scarcity of resource what are the preventive measures should be taken and give your ideas?

5. Distinguish environmental aspects of conventional and non-conventional energy technologies?

6. Explain in detail the environmental aspects of non renewable energy technology?

7. What is global warming and explain their effects on the environment?

8. What are the impacts on environment by burning fossil based fuels and mention their causes and effects.

9. Describe three things that you can do practically in your life today to reduce your energy footprint.

10. What is an ecological footprint with an example explain how do you assess the carbon footprint of person?

11. Can we afford to wait for the end of the crises, that follow one after the other, before addressing the energy problem and the related climate change

12. What do you understand by the terms carbon footprint and ecological foot print? Explain how each individual person is creating carbon footprint? As a technocrat how can you reduce your and your family carbon footprint?

13. Which one will have more carbon footprint among the milk imbibe directly from cow and the milk obtained from dairy? Judge your answer with suitable reasons?

14. What is a carbon foot print? How can you estimate the carbon footprint of an individual?

15. Generalize the environmental impacts of solar and wind energy?

16. A severe threat to environment by using coal as a source for electricity. Mention those severe environmental impacts and explain them?

17. What do you understand by the term bio capacity?

Objective type questions

1. The pollutant responsible for ozone holes is _____

   a) CO_2 b) SO_2 c) CO   d) CFC

2. Which of the following is a major pollutant causing acid rain?

   a) CO_2 b) Sulphur dioxide   c) Hydrogen peroxide
   d) Carbon monoxide

3. A natural phenomenon that becomes harmful due to
   pollution is _____

   a) Global warming   b) Ecological balance

   c) Greenhouse effect   d) Desertification

4. The radiation in the sunlight that gives us the feeling of
   hotness is _____

   a) Visible radiation b) infra-red   c) Red   d) ultra-violet

5. Harmful radiation emitted by the sun is _____

   a) Visible   b) infra-red   c) Ultra-violet   d) radio waves

6. Which of the following causes the least pollution when
   burnt?

   a) Petrol   b) Diesel   c) Coal   d) Natural gas

7. The average earth's temperature is about---------

   a) 28°C   b) 15°C   c) 32°C d) 10°C

8. The unit of Carbon foot print is

   a) Meters   b) Number of pages/year
   c)Tons/year   d)Hectares/year

9. Which of the following is not a major green house gas?

   a) CO_2 b) Calcium carbonate c) water vapor   d) methane

10. Which of the following problems is associated with the burning of coal?

a) Acid rain   b) CO_2 emissions

c) Ash with toxic material emission   d) all of the above

11. Which among the following does not produce carbon dioxide emissions?

a) Oil   b) Uranium   c) coal   d) Natural gas

12. Greenhouse effect refers to increase in

a) Global temperature.   b) Carbon monoxide

c) atmospheric pressure   d) Greenery

13. The Ozone layer in the atmosphere acts to protect

a) X-Rays   b) UV-A Rays   c) UV-B Rays   d) Infra red Rays

14. Which of the following does not damage ozone layer?

a) CFC   b) CCl_3 c) HCFC   d) HFC

15. Ozone has

a) Three oxygen atoms   b) Two oxygen atoms

c) Two chlorine atoms   d) Two bromine atoms

16. The burning of fossil fuels does not release:

a) Potential energy   b) Light energy c) Sound energy d) Heat energy

17. The main constituent of CNG is _____

a) Methane   b) butane   c) ethane   d) Propane

This Chapter deals with the supply chains consist of three main phases like production, distribution, and disposal. Between the phases of distribution and disposal is the product usage, but since that phase is directly controlled by the consumer it is excluded from the supply chain point of view. Eventually, we must build a green index that reflects all these three phases of a product life-cycle in a supply chain. This index can then be used variously for designing green supply chains across industries as a sustainability measure that goes across industries, corporate supply chains, and consumers to create sustainable business practices.

5.1 Supply Chain

1. **Energy profile (Production).** The energy profile will model the total energy requirements of producing the raw materials as well as the manufacturing process that converts them into finished merchandise. Such data will be typically supplied by the manufacturers, through a process very similar to the product specifications that the manufacturers provide today. The existing data pools like GDSN may be expanded to include this data for the energy profiles of the manufacturing process and the energy profiles of the raw materials.

2. **Distribution profile.** This will capture the carbon footprint of the material movements required to manufacture a given product, with elements such as the distances traveled by the raw materials from their source to the factories, and by the finished goods to reach the retailer's warehouses and stores from the factories. The modes available on these routes and

energy profiles of these modes may affect such scores. It may also capture the distribution unit profile based on packaging that affects distribution costs.

3. **Recycle profile (Disposal).** This profile will model the material's recycling characteristics, types of facilities required, and regional laws governing recycling requirements by collecting data on the recycling profiles for the merchandise as well as for the packaging materials.

Every stage in the electrical energy supply chain provides is an opportunity to implement the technologies and to save energy.

Energy Supply chain of Thermal power plant:

Coal mines ⟹ Pulverizer ⟹ Boiler ⟹ Heat exchanger ⟹ Steam ⟹ Turbines ⟹ electricity End Use (Industries, Agriculture, Transportation and Domestic).

Energy Supply chain of Oil (Petrol and Diesel):

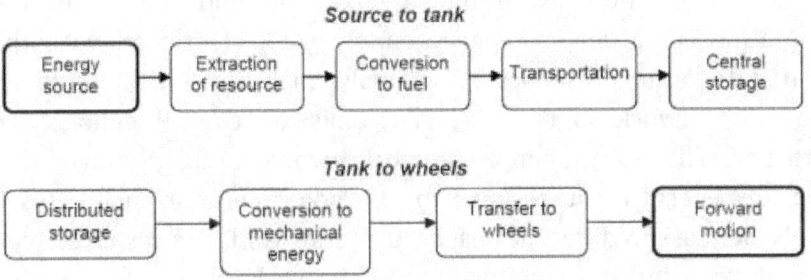

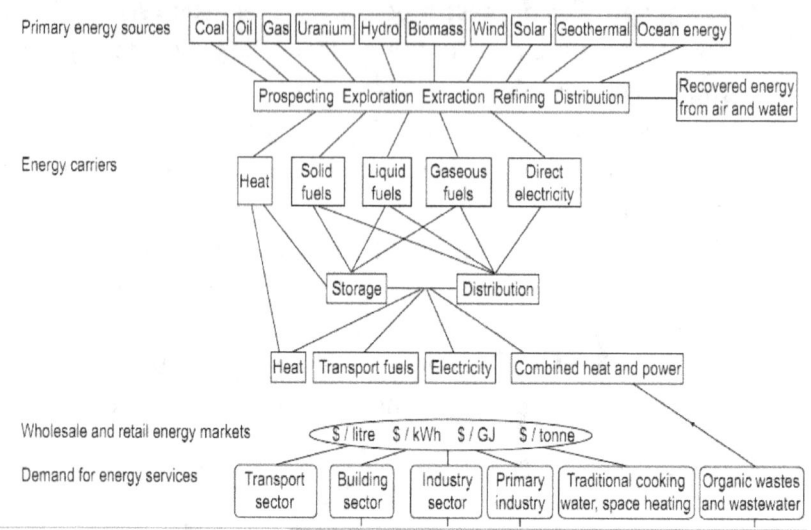

Figure 5.1: Energy Supply Chain *(Source: Green Rhino Energy)*

Energy supply chains aim to ensure an uninterrupted satisfaction of energy demand at various forms with the maximum efficiency, while satisfying constraints on emissions released to the environment. The present work offers a framework for the optimization of the operating conditions in energy supply chains enriched with the option to optimally allocate new polygeneration units within an existing network so that stringent emission control policies are satisfied. The energy supply network involves a supply layer that includes energy resources available for polygeneration such as fossil fuels, biomass in different forms, solar and wind. The resources can be utilized by the energy supply network at various energy polygeneration units that include gas turbines, boilers and steam turbines, PV panels, wind generators and so forth. Within each unit, energy can be converted in various forms (e.g., electric power, steam of variable quality,) dictated by the overall demand and the decisions taken at the optimization level. The behavior of the system is monitored over a time horizon that enables the feasibility of the network to achieve the production quotas and satisfy the emission requirements. The decision support framework enables the optimal operation through structural changes in the network and assignment of suitable production targets for the individual units.

Energy supply chains aim to ensure an uninterrupted satisfaction of energy demand at various forms with the maximum efficiency, while satisfying constraints on emissions released to the environment (Klemes et al, 2005). To this extent, energy supply chains have incorporated various energy resources including biomass and other renewable energy sources to traditional and conventional resources such as coal, natural gas and other fossil fuels in order to reduce hazardous pollutants and other greenhouse gases. Numerous approaches investigate the spatial allocation of biomass and other renewable energy sources units for the improved performance of existing energy supply chains (Lam et al., 2009 and 2010; Leduc et al., 2010). Other research works focus on the design aspects of polygeneration units that are mainly governed by the strong interaction of chemical synthesis and power generation (Liu et al., 2010; Pistikopoulos et al., 2010). The efficiency of process systems engineering tools in the synthesis and design of energy systems and in particular biorefineries is investigated by Kokossis and Yang, 2010. The current approach aims at determining critical decisions in the performance of an energy supply chain both at the strategic and the operating level. Energy supply chain is considered as the distributed production of power, heat and chemicals by a set of fixed polygeneration units utilizing different resources. The strategic level involves decisions regarding the introduction of new energy polygeneration units into an existing infrastructure through targeted investments. The outlined decisions involve the operating conditions and participation in the generation of energy products of existing or newly introduced units in an annual basis for the satisfaction of the demand for power, heat (in the form of steam at different pressure levels and chemicals) under significant variability. The proposed optimization framework enables the identification of demand driven highly performing solutions through the direct consideration of all interacting factors within the energy supply chain over a future time horizon.

Optimization framework for energy supply chains

The present work offers a framework for the optimization of the operating conditions in energy supply chains enriched with the option to optimally allocate in time new polygeneration units within the energy supply network so that stringent demand profiles and emission control policies are fully satisfied. The energy supply network involves a resources layer that includes energy resources available for polygeneration such as fossil fuels, biomass in different forms (e.g., pellets of various biomass stocks, gasification of lignocellulosic residues), solar, wind, geothermal and hydro energy resources. The resources can be utilized by the energy supply network at various energy polygeneration units that include gas turbines, boilers and steam turbines, PV panels, wind generators and so forth. Within each polygeneration unit, energy can be converted in various forms (e.g., electric power, steam of variable quality, cooling, chemicals) as dictated by the overall demand profiles and the decisions taken at the optimization level.

The demand of the various forms of energy behaves stochastically and is modeled with a non-stationary time series model. Each polygeneration unit produces as side-products emissions (e.g., greenhouse emissions, NOx) that are monitored by regulating agencies and local legislation. A schematic of the super-structure of an energy supply chain under consideration is shown in Figure 5.1.

The optimization of the supply chain is performed at two different modes:

(a) The operational level, where the operating decisions regarding the distribution of the energy products (including chemicals) coming out of each individual unit and the overall utilization of the polygeneration unit in the distributed production scheme is determined and (b) the strategic level, where decisions about the incorporation of new Polygeneration units to an existing energy supply chain are taken. In the operational level, variability in the supply and the prices of the raw materials and the energy demand are taken into direct consideration for a fixed structure of the energy conversion network and emission constraints for a time horizon extended into the future. The time horizon enables the feasibility of the network to achieve the production quotas and satisfy the emission requirements. In the strategic level, the decision periods become larger than in the operating level study but for a flexible network structure. The main objective is therefore the identification of suitable conditions for the introduction into an existing energy supply chain in terms of economic benefit of energy conversion units

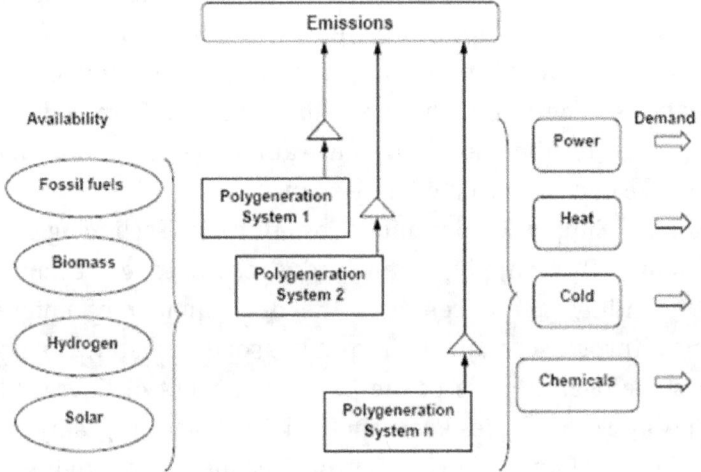

that utilize renewable energy sources.

Figure 5.2: General schematic energy supply chain for renewable and non renewable source

The prediction of the behaviour of the polygeneration infrastructure is achieved through the employment of simplified models. The characteristics of the various polygeneration systems at different operating modes are studied using the rigorous process simulator Aspen Plus (www.aspentech.com). Regressed equations are then derived using the predictions of the process simulator, thus relating the heat and power generation quantities with the required raw materials for the capacity range and the operating modes of interest. Similarly, CO_2 capture plants based on CO_2 absorption by suitable amine blends are considered. The introduction of additional more complex polygeneration units is therefore straightforward as the required modelling information can be derived from validated process model simulations. Estimates of the investment and the operating costs for new and existing units enable the rigorous evaluation of alternative design options. The optimization is performed over a future time horizon so that the effects of decisions at an early time can be assessed. For instance, the total capacity for new units introduced into the system is not based only on current needs but also on future needs both in demand and emissions specifications. Therefore, the decisions are not obtained with a myopic view of the situation but rather based on an overall performance over the selected time span.

Photovoltaic Supply Chain and Cross-Cutting Technologies: The Photovoltaic (PV) Supply Chain and Cross-Cutting Technologies project identifies and accelerates the development of unique PV products or processes that will impact the solar industry. The project supports the overall goals of the U.S. Department of Energy (DOE) Solar Energy Technologies Program (SETP or Solar Program).

The PV Supply Chain and Cross-Cutting Technologies activities are a component of the systems development and manufacturing activities, within the SETP PV subprogram

Green Supply Chains: Beyond the Cost of Energy

In order to reduce operating costs and achieve sustainable development of competitive advantage, modern enterprises pay attention to supply chain system. At the same time, the worldwide environmental awareness introduces the concept of environmental protection into the supply chain management system. Therefore, the green supply chain management has been concerned by academia circles and business circles from home and abroad. Green Supply

Chain Management is considered to strengthen environmental awareness and recycle resourced in the supply chain management, through the upstream and downstream enterprises' cooperation, and all sectors of the enterprise communication, including product design, material selection, product manufacturing, product sales and the recovery. All the process should be considered to optimize the environmental efficiency as a whole and achieve the sustainable development of the enterprises and its supply chain.

5.2 Institutional Aspects

Institutions are defined as 'the humanly devised constraints that shape human interactions'. More informally, they are 'the rules of the game', while the players of the game are the 'organizations'. The costs of reducing uncertainty in human interactions are fairly dependent on the quality of the institutions, and how seriously these are enforced.

Importance of these aspects:

The *World Bank* considers land titling, land registration, funding, and information supply in general as examples of institutional development. *(Feder & Feeney, 1991)* place property rights in general and land rights in particular in the context of institutional structure of society and economy. They distinguish three basic categories of institutions, namely the constitutional order, institutional arrangements, and normative behavioral codes. The constitutional order refers to the fundamental rules about how society is organized: the rules for making rules. Institutional arrangements include laws, regulations, and (*inter alia*) property rights. The normative behavioral code refers to the values that legitimize the arrangements. They conclude that property rights are an important class of institutional arrangements, as 'property 'implies a system of relations between individuals, by creating mechanisms for the definition and enforcement of these rights including both formal procedures and social customs and attitudes concerning the legitimacy and recognition of those rights

- Local and business community opposition to the removal of/ restrictions on spaces for energy;

- Availability and acquisition of right-of-way or physical space; Integration of multiple priorities, objectives, and agendas

- Concerns over long term funding commitments to energy;

- Impacts of energy on environment operations;

- Finding political champions to support energy projects

- Gaining community support for sustainable development;

- Educating the public on energy use;

- Managing perceptions and expectations.

Institutional Considerations

Institutional aspects of policy, planning, management, financing, service infrastructure, community participation, and user interface are critical to the successful use of any chosen technology. While focusing mostly on various institutional, organizational, and social issues that should be addressed prior to initiating a large scale project, much of this information is applicable to smaller scale projects as well.

Policy and Commitment

A national policy framework is key to the integration of RE into health clinic electrification. Without an established policy at the highest level and commitment at every other level, efforts to implement RE technologies run a high risk of failure. Many health professionals responsible for allocation of limited financial and human resources for public health programs see RE technologies as costly, experimental technologies with a history of failures. They may be unaware of the benefits of successful applications. Promotion of RE development should be accompanied by the removal of barriers such as import duties and subsidies for fossil fuel power. Partnerships offer the ability to share resources, costs and benefits. Collaboration with other community sectors such as education and agriculture is beneficial to the mission of public health programs.

Similar needs such as light for education and safe community water supply can share the same service and share its costs. Collaboration between public and private entities should seek to support public health with private commercial services.

Decision-Making and Management

In some instances RE systems failed to prove reliable. These experiences have often left the perception that RE is more costly and requires special skills and knowledge. Without quality standards and proper service these failures will be repeated. The selection and procurement of quality systems and service must be based upon standards of acceptance and quality control. Adequate resources for logistics, servicing, and maintenance must be allocated.

Higher quality service is the strongest argument for integrating RE into health care. Cost analysis is usually based upon the initial cost. LCC analysis can be a favorable argument for renewable energy choices but is not a determinant.

Service Infrastructure

Although the technologies are mature, there lacks a well-developed, reliable service industry to install, maintain, and repair systems in the field. This infrastructure can be developed more cost effectively through partnerships with other human service sectors to share costs. System designers and field technicians must receive regular training and information to retain current knowledge and skills. There must be enough pay for their services to keep them available and prepared. Spare parts must be available. The system users must be aware of the proper operation and care for the system. Training must be adequate and regular.

Financing and Income Generation

Availability of financing for purchase of equipment will put the benefits of RE technologies within the reach of those that need them most. National governments and international donors have provided significant capital funds for clinic electrification with marginal results. The lack of operating funds for maintenance and repair soon renders the equipment unusable. There are numerous examples of revolving credit funds to support productive use enterprises and enable individuals to purchase solar lighting systems as a means of rural electrification. The principles of these successes should be applied to community-based health facilities and enterprises. Strategies to generate income at the local level can enable recovery of the loan funds and provide operating funds to maintain the equipment.

Community and Culture

Dynamics within the community play an important role in making or breaking the success of a project. The community and its individuals may support, ignore, abandon, or even subvert the system. Systems and services provided must be in the appropriate language and level of education to be understood

The one million dollar concentrating solar power program provides technical assistance to the Government of India to overcome barriers that hinder the deployment of solar technology. The barriers include :

- ✓ Lack of ground validated solar radiation data,
- ✓ Performance uncertainties,
- ✓ Undeveloped CSP industrial base,
- ✓ Continuing inequities in public subsidies of energy technologies, and
- ✓ Inadequate policy frameworks especially in developing country context.

The World Bank has had a long engagement in developing renewable energy resources, and has assisted partner countries in their efforts to address issues related to creating a favorable environment for new and proven renewable energy technologies. Bank experience and expertise in policy development and leveraging financial recourses will help set up enabling provisions to accelerate concentrating solar power deployment in India.

To-date, numerous activities have been completed to further concentrating solar power development in India such as:

- ✓ The recent support to the Delhi International Renewable Energy Conference (DIREC) where knowledge about domestic and international practices, new technologies, policy and regulatory developments in solar thermal sector were disseminated.
- ✓ Completion of a study on the local capabilities of Indian manufacturers in developing CSPs.
- ✓ A study on barriers to solar power development in India to provide insights on next steps for accelerated development of solar power in India.
- ✓ Study tour for Indian policy makers and regulators to Spain in June 2010 to gain insight on two main solar thermal technologies, namely, solar tower technology and parabolic trough.
- ✓ Participation of practitioners, policy makers, and regulators in several CSP international conferences.

By 2012, the government of India expects that renewable energy will contribute about 10 percent to the total power generation capacity of India. The success of Concentrating Solar Power deployment heavily depends on a favorable regulatory environment that would provide a range of various incentives to allow Concentrating Solar Power installations competing with conventional power plants.

1. Policies, strategies and projects

The reasons for the positive development in the past lie in the fact that RE has been a long-term priority of Austria's energy policy, resulting in a broad range of subsidies and research facilities. This has been a result of the goal to promote the development of RE early on in order to ensure an environmentally sound sustainable energy supply. Moreover, the policy of the Austrian Agricultural Ministry emphasises three priorities:

- ✓ Improving product quality
- ✓ Fostering sustainable agriculture
- ✓ Preserving family farms

2. The role of technical agencies

Government Organizations involved:

Ministry of Agriculture and Forestry: supports SRF production, establishment of rural district heating plants, purchase of biomass heating system, research, financial support;

Ministry for Economic Affairs: energy policy, energy statistics

Ministry of Environment: CO_2 reduction, Kyoto issues, support of clean technologies

Ministry for Science and Traffic: Research, Universities

On a provincial level, the departments of the provincial governments such as the Energy Commissioners of the provinces and the Energy Efficiency Agencies

Non-Government Organizations involved:

Biomass Association: shapes the national and international public opinion about environmental and energy policy and inform decision makers in politics about the possibilities and the advantages of biomass;

Institute for Agricultural Engineering, a research and development, trial and training centre concerned with agricultural mechanisation, bioenergy, and renewable resources;

research organisations specialises in research on renewable energy, energy conservation, and emissions from energy conversion processes. The institute works in co-operation with industry, government organisations, and service enterprises such as heating plant operators.

3. Legal issues

Minimal efficiency and emission limitations for automatic residential heating systems up to 350kW are determined as a general rule in agreements between the State Govt and the provinces.

According to the Law, the targets of environmental planning are to be determined in general. The provinces and municipalities shall formulate in detail how they will be carried out such targets.

Financial Regulations: State and provincial governments are subsidising almost all RE-based technologies. The public funding available for research on RE has been rising continually over the last few years

State Government's subsidies:

Provincial Governments subsidies:

- local Biomass fired heat distribution is sponsored in Styria, particularly when the initiative comes from a group of farmers and foresters;

- Biomass fired heat distribution is subsidised

- Biomass plants, district heating systems, and research are subsidised

Financial Incentives for the Production of Energy Crops:

- direct investment subsidies are available

- Subsidised agricultural investment loans are granted for a period of approximately 10 years.

Financial Incentives for the Conversion of Energy Crops:

- Subsidies up to 40 percent of the investment cost are available for Biogas facilities.

Financial Incentives for the Utilization of energy from Energy Crops:

- biomass district heating systems by farmers co-operatives: subsidies up to 50 percent;

- district heating systems based on wood or straw: subsidies up to 40 percent of the investment cost;

- automatic wood firing systems for individual houses: subsidy;

- Electricity produced from RE: agreement under negotiation - utility companies will be required to buy electricity produced at special rates of 20-45 percent above the normal tariff.

5.3 Energy Usage

The usage of energy was divided in different sectors as the following table shows.

Residential consumption (domestic):	10%
Transport (buses, trains, trucks, planes & ships):	22%
Agricultural consumption:	7%
Industry (manufacturing sector):	49%
Others (hotels, hospitals, offices, shops etc):	12%
Total:	100%

Finally the energy produced from the production plant has reached to the society with having different losses while the distribution to the various sectors which were pellucidly given below.

60% of the heat energy is wasted before it gets converted into electricity.

10% of the produced electrical energy is lost in transmission.

10% of the remaining is lost in distribution.

20% of the energy is wasted due to either indifferent attitude or theft.

Global energy consumption has increased steadily for much of the twentieth century, particularly since 1950. In 2008, the world consumed 11294.9 million tons (Mtoe) of oil in total energy (including hydroelectric and nuclear power). That amount corresponds to over 82 billion barrels of crude oil burned in 2007.2009 saw the first energy consumption decline since 1982, presumably due to the global recession. Total energy consumption was lower in 2009 than in 2008, but current 2010 data show consumption rates rising again.

One of the most notable developments in recent years has been the explosion of growth in energy demand from Asia, which eclipsed North America for the first time in 2003 as the world's most energy hungry region. Much of this increase in demand came from China and India. As China is rapidly industrializing, its need for energy is constantly growing. In the last few years, China has catapulted itself into second place for energy consumption behind the United States in global rankings. The next several largest consumers, Russia, Japan, India, Germany, Canada, France, the United Kingdom, and Ukraine, come in distantly behind in the rankings. Although China is still behind the United States in terms of energy consumption, its total consumption is increasing at breakneck speed.

With respect to oil, for example, China's share of global consumption is only approximately 9.6 percent *(est. 2008, source BP)*. The World Energy Outlook 2008 concluded, however, that from 2006 to 2030, China and India will account for over half of the world's increase in annual energy demand. U.S. demand is projected to slowly decrease. The slow decline will result from continued high oil prices and recent legislation enacted on energy efficiency. Even where the percentages may differ, the absolute numbers describing Chinese demand growth for many types of energy easily rival those of the United States. Total consumption is only one measurement of energy usage. Others, such as per capita energy consumption and energy intensity, offer more nuanced information about differences among countries.

Per capita energy consumption has remained relatively stable since 1980 both worldwide and in the United States. This indicates that, although the world's total consumption has increased, most individuals in most countries use about the same amount of energy they did 20 years ago. From this, we learn that much of the increase in total demand can therefore be attributed to population growth and social transformation e.g. the integration of millions of people into modern, urbanized communities in China and India.

The electricity sector in India had an installed capacity of 185.5 GW as of November 2011, the world's fifth largest. Thermal power plants constitute 65% of the installed capacity, hydroelectric about 21% and rest being a combination of wind, small hydro, biomass, waste-to-electricity, and nuclear. The per capita average annual domestic electricity consumption in India in 2009 was 96 kWh in rural areas and 288 kWh in urban areas for those with access to electricity. India currently suffers from a major shortage of electricity generation capacity, even though it is the world's fourth largest energy consumer after United States, China and Russia.

The International Energy Agency estimates India needs an investment of at least $135 billion to provide universal access of electricity to its population. Of the 1.4 billion people of the world who have no access to electricity in the world, India accounts for over 300 million. The International Energy Agency estimates India will add between 600 GW to 1200 GW of additional new power generation capacity before 2050.

Total world energy supply i.e. primary energy (2008) was 143,851 TWh and the end use of energy (2008) was 98,022 TWh. The difference 32 % is energy losses. Energy losses are not constant but depend on the energy source and technology.

S. No	Country	Per capita of energy consumption in kWh/year as of 2003 data
1.	India	682
2.	Pakistan	608
3.	Bangladesh	214
4.	China	1516
5.	Russia	5890
6.	New Zealand	5831
7.	Singapore	6870
8.	Germany	5597
9.	Australia	7622
10.	United Kingdom	5218
11.	United States	10381
12.	Qatar	28495

Table 5.1 Worldwide Per capita of Energy Consumption in kWh/year

(Source: This data published by World resources Institute for the year 2003)

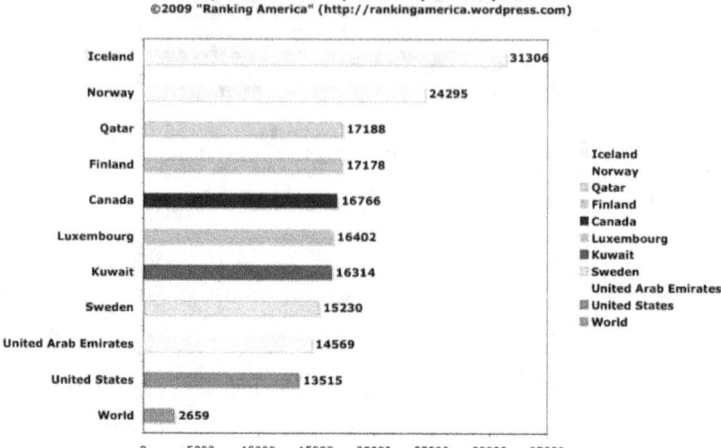

Figure 5.3 Top countries by electricity consumption
(Source of this figure: rankingamerica.worldpress.com)

Review Questions

1. Define an institution and narrate why institutional aspects are important for energy sector?

2. What is supply chain? How is it differing from green supply chain?

3. What is energy supply chain? Draw an energy supply chain diagram for renewable energy systems.

4. Explain the different phases of energy supply chain with an example?

5. Draw schematic layout of energy supply chain and explain energy flow of a thermal power station?

6. Write a note on energy usage in Indian society?

7. Write a short note on energy and per capita of energy consumption in India?

8. What is role play of energy Institutions in India and explain their importance?

9. Write institutional aspects of energy and discuss the importance of these aspects.

10. Write about energy supply chain (from producer to consumer).

11. Compare the consumption and production of electricity in India and the World?

12. Give the contributions of energy use by different sectors in India and give some examples of uses of different forms of energy in our daily life.

CHAPTER 6

EFFICIENCY OF ENERGY SYSTEMS

This chapter mainly analyzing the efficiency of energy systems in the various production plants and also in the domestic appliances; which will enlightens us the effective usage of different appliances in our domestic purpose without having any losses by upgrading with new equipments and creating awareness among the people about the conservation of energy.

6.1 Energy Efficiency

Residential/Commercial:

Much of this energy is not put to use. Heat pours out of homes through drafty doors and windows, as well as through ceilings and walls that aren't insulated. Some appliances use energy 24 hours a day, even when they are turned off. Energy-efficient improvements can make a home more comfortable and save money. Many utility companies provide energy audits to identify areas where homes are wasting energy. These audits may be free or low cost.

Lighting:

Energy conservation is any behavior that results in the use of less energy. Energy efficiency is the use of technology that requires less energy to perform the same function. A compact fluorescent light bulb that uses less energy than an incandescent bulb to produce the same amount of light is an example of energy efficiency. The decision to replace an incandescent light bulb with a compact fluorescent is an example of energy conservation.

Incandescent lamp: Incandescent bulbs do this by shooting electricity into a thin metal filament surrounded by inert gas and encased in a glass shell. Metal normally emits invisible infrared light when heated like this, but get the atoms worked up enough and they'll produce a visible glow, too.

Compact Fluorescent light:

Metallic atoms are also the light source in fluorescent lamps, but they use vaporized mercury instead of a solid filament. The incoming electrical current is carried through a glass tube, straight or coiled, that's filled with mercury vapor and argon gas. The electrified mercury atoms begin vibrating and releasing invisible ultraviolet light, which in turn excites a fluorescent phosphor coating on the inside of the tube, finally producing the visible light.

If you replace 25 percent of your light bulbs with fluorescents, you can save about 50 percent on your lighting bill. Compact fluorescent light bulbs (CFLs) provide the same amount of light and do not flicker or buzz. CFLs cost more to buy, but they save money in the long run because they use only one-quarter the energy of incandescent bulbs and last 8-12 times longer. Each CFL you install can save you $30 to $60 over the bulb's life.

Some light bulbs are better than others for the environment, and in order to find out which ones are better, simply compare them by how much energy they need to produce light. Both halogen and incandescent bulbs produce light by heating a tungsten filament with an electrical current. Compact fluorescent lamps (CFLs), on the other hand, create light through an entirely different mechanism. The fluorescent gas inside the bulb produces ultraviolet light when electrified, and the lamp's coating converts the ultraviolet light into visible light. Because of this, CFLs are between 67 percent and 80 percent more energy-efficient than incandescent bulbs. Halogen lamps stand somewhere in between, ranked as more efficient than normal incandescent bulbs, but not as efficient as fluorescent lights.

Energy Usage Efficiency:

The following example shows the inefficiencies involved in converting a primary energy supply into useful light output. A typical 60 Watt incandescent lamp produces illumination of about 15 lumens per Watt of applied power. The total light output from the bulb is therefore 900 lumens, which is equivalent to about 1.35 Watts or 1.35 Joules per second of radiated light power, and the conversion efficiency is 2.25%. The rest of the applied electrical

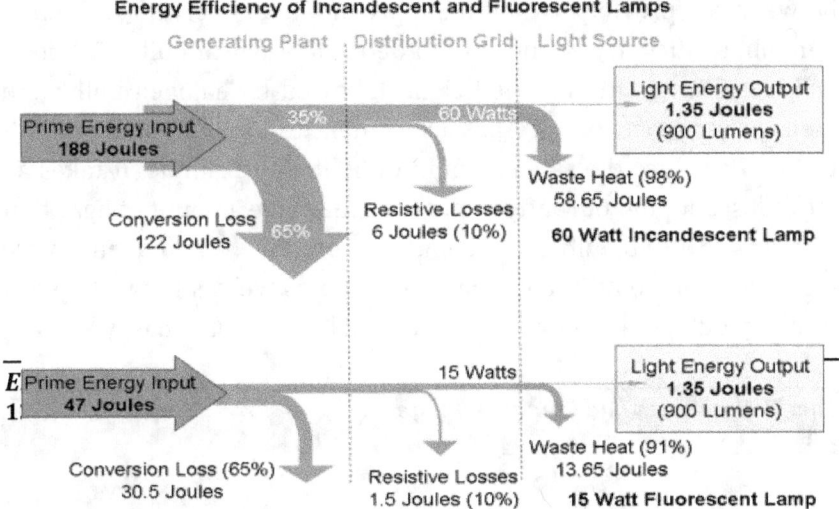

Energy Efficiency of Incandescent and Fluorescent Lamps

energy is lost as heat. Taking into account the typical efficiency of electricity generating plant of 35% and 10% Joule heating losses in the distribution grid, the efficiency of converting primary energy into light energy is only 0.7%

Figure 6.1 Losses in the Electric Lamp *(Source: mpoweruk/electropaedia/energy efficiency)*

For comparison, a compact fluorescent lamp (CFL) produces between 50 and 60 lumens per Watt. By using fluorescent rather than incandescent lamps, the power consumption of the lamps can be reduced from 60 Watts to 15 Watts for the same light output. The consumer saves a modest 45 Joules per second but the corresponding prime energy consumption is reduced by a massive 141 Joules per second.

Luminous Efficiency

Light is measured in units called "lumens," which correspond to the amount of light produced per watt. For a source of light to be 100 percent efficient, it would hypothetically need to give 680 lumens per watt. The luminous efficiency of fluorescent lighting is the highest, between 9 percent and 11 percent for most CFLs, while conventional incandescent bulbs stand between 1.9 percent and 2.6 percent efficiency. The luminous efficiency of halogen lamps cradles between the previous two at an approximate 3.5 percent efficiency. Luminous efficiency is one way to determine which bulb to choose, yielding CFLs as the most efficient, followed by halogen bulbs and then incandescent bulbs. Another element to look at is the watts it takes to produce the same amount of light. For example: It takes an incandescent bulb 60 watts to produce the same amount of light that would take a CFL bulb only 15 watts to produce. Manufacturers are required to list both the lumens produced as well as the watts used by every bulb, so luminous efficiency can be calculated easily.

Halogen Bulbs vs. Conventional Incandescent Bulbs

Incandescent bulbs, including halogen bulbs, produce light by heating a filament of tungsten metal until it is white hot. In a normal incandescent bulb, the tungsten slowly vaporizes and deposits on the inside of the bulb until it is too thin to carry an electric current and the bulb burns out. Halogen bulbs are filled with a special gas that causes the vaporized tungsten to be deposited back onto the filament instead of the inside of the bulb. Halogen bulbs last longer and also burn hotter than conventional incandescent bulbs, making them slightly more efficient. However, these gains may be negated by the extra energy an air conditioner must use to cool a room.

New Energy-Efficient Halogen Bulbs

Another energy-saving choice, halogen bulbs are incandescents that use halogen gas around the filament; this increases the efficiency and lifespan of the bulb. Halogens use only about 25 percent less energy than a traditional incandescents, however, so are a distant third behind LEDs and CFLs in energy efficiency.

A new class of halogen bulbs has recently been developed. These new bulbs use a special infrared coating to redirect infrared light back toward the tungsten filament, reducing waste heat and improving efficiency by up to 30 percent over typical incandescent bulbs. They are still not as efficient as CFLs, which are around 75 percent more efficient than normal bulbs, but this variety offers top-tier efficiency when it comes to halogen bulbs.

LEDs:

According to Energy Star, as of 2011, LED bulbs use only 20 to 25 percent of the energy of incandescent bulbs, with efficiency matching or exceeding fluorescent lights. In addition to energy efficiency, a U.S. Department of Energy report notes that the "benefits of LEDs include long operating lifetime, lower maintenance and life-cycle costs, reduced radiated heat, minimal light loss, dimmability, controllability, directional illumination, durability, safety improvements, and light pollution reduction."

The Energy losses in a Plant

Much of the energy content of the available energy sources is wasted by inefficiencies the energy conversion and distribution processes. Considering domestic electric lighting as a typical example, less than 1% of the energy consumed to provide the electricity is ultimately converted into light energy. The other 99% is wasted in the supply chain. Using conventional fossil fuelled generating plant, losses accumulate as follows:

- 10% of the energy content of the fuel is lost in combustion and only 90% of the calorific content is transferred to the steam.
- The steam turbine efficiency in converting the energy content of the steam into mechanical energy is limited to about 40%.
- The rotary electrical generator is very efficient by comparison. The conversion efficiency of a large machine can be as high as 98% or 99%.
- Transmission of the electrical energy over the distribution grid between the power station and the consumer results in a distribution loss of 10% mainly due to the resistance of the electrical cables.
- Further energy is lost due to the energy conversion efficiency of the end user's appliance. Incandescent lighting is particularly inefficient converting only 2% of the electrical energy into light.

The efficiency of electricity generating plant fuelled by biomass is typically only around 35%

Hydroelectric Power Generation Efficiency

Hydroelectric power generation is by far the most efficient method of large scale electric power generation. Energy flows are concentrated and can be controlled. The conversion process captures kinetic energy and converts it directly into electric energy. There are no inefficient intermediate thermodynamic or chemical processes and no heat losses.

The conversion efficiency of a hydroelectric power plant depends mainly on the type of water turbine employed and can be as high as 95% for large installations. Smaller plants with output powers less than 5 MW may have efficiencies between 80 and 85 %. It is however difficult to extract power from low flow rates.

Practical Power and Conversion Efficiency

German aerodynamicist Albert Betz showed that a maximum of only 59.3% of the theoretical power can be extracted from the wind, no matter how good the wind turbine is, otherwise the wind would stop when it hit the blades. He demonstrated mathematically that the optimum occurs when the rotor reduces the wind speed by one third. After inefficiencies in the design and frictional losses are taken into account the practical power available from the wind will rarely exceed 40% of the theoretical power.

Converting this wind power into electrical power incurs further losses of 10% or more in the drive train and the generator and another 10% in the inverter and cabling such that ultimately, the wind turbine will capture only about 30% to 35% of the wind energy available

Electricity Distribution Efficiency

The demand for electrical energy is ever increasing. Today over 21% (theft apart) of the total electrical energy generated in India is lost in transmission (i.e., 4-6%) and distribution (i.e., 15-18%). The electrical power deficit in the country is currently about 18%. Clearly, reduction in distribution losses can reduce this deficit significantly. It is possible to bring down the distribution losses to a 6-8 % level in India with the help of newer technological options (including information technology) in the electrical power distribution sector which will enable better monitoring and control.

Distribution Loss Factors (DLF)

The resistance of the cables conducting the current flow between the generating plant and the end user's premises cause further efficiency losses due to the Joule heating (I^2R Losses) of the interconnecting power cables. There are two major influencing factors.

- **Location:** The resistance of the cables increases with distance so that losses are typically 5% for supplies to urban locations close to the power source but as high as 10% to 20% for remote rural locations. The overall average for the USA is 7% to 8%.

- **Voltage:** Since Joule heating losses are proportional to the square of the current, distribution losses can be reduced by transmitting the power with as low a current as possible by using higher transmission voltages. The upper voltage limit is set by the breakdown of the air insulation between the power cables and the earth, or more likely across the insulators suspending the cables from the transmission pylons (towers).

With high voltage transmission systems there are also additional, though minor, copper and iron losses in the transformers, stepping up the voltage at the generating station and stepping it down again at the point of consumption, due to the resistance of the windings and the hysteresis and eddy current losses in the transformer cores.

Earning the ENERGY STAR means products meet strict energy efficiency guidelines set by the US Environmental Protection Agency. Lighting products that have earned the ENERGY STAR deliver exceptional features, while using less energy. Saving energy helps you save money on utility bills and protect the environment by reducing greenhouse gas emissions in the fight against climate change.

An ENERGY STAR qualified light bulb:

- Saves money about $6 a year in electricity costs and can save more than $40 over its lifetime

- Meets strict performance requirements that are tested and certified by a third party

- Uses about 75% less energy than a traditional incandescent bulb and lasts at least 6 times longer

- Produces about 75% less heat, so it's safer to operate and can cut energy costs associated with home cooling

Effects on Climate

Compared to incandescent bulbs, fluorescent lamps are especially efficient in warm climates. Around 90 percent of the energy used to power an incandescent bulb is transformed into heat, as opposed to 30 percent for CFLs, which use less electricity to begin with. When it's hot out, switching to CFLs not only reduces electricity for lighting, but it also reduces workloads on air conditioners. The opposite is true in cold climates. Without the extra heating from incandescent bulbs, more natural gas or oil needs to be burned to heat homes and businesses. In areas where electricity is cheap or comes from non-fossil fuel sources, switching to CFLs can actually increase overall energy costs and greenhouse gas emissions.

Heating and Cooling:

Heating and cooling systems use more energy than any other systems in our homes. Typically, 44 percent of an average family's energy bills is spent to keep homes at a comfortable temperature. You can save energy and money by installing insulation, maintaining and upgrading the equipment, and practicing energy-efficient behaviors. A two-degree adjustment to your thermostat setting (lower in winter, higher in summer) can lower heating bills by four percent and prevent 500 pounds of carbon dioxide from entering the atmosphere each year. Programmable thermostats can automatically control temperature for time of day and season.

Insulation and Weatherization:

You can reduce heating and cooling needs by investing in insulation and weatherization products. Warm air leaking into your home in summer and out of your home in winter can waste a lot of energy. Insulation wraps your house in a nice warm blanket, but air can still leak in or out through small cracks. Often the effect of small leaks is the same as keeping a door wide open. One of the easiest money-saving measures you can do is caulk, seal, and weather-strip all the cracks to the outside. You can save 10 percent or more on your energy bill by stopping the air leaks in your home.

Doors and Windows:

About one-third of a typical home's heat loss occurs through the doors and windows. Energy-efficient doors are insulated and seal tightly to prevent air from leaking through or around them. If your doors are in good shape and you don't want to replace them, make sure they seal tightly and have door sweeps at the bottom to prevent air leaks. Installing insulated storm doors provides an additional barrier to leaking air. Most homes have many more windows than doors.

Replacing older windows with new energy-efficient ones can reduce air leaks and utility bills. The best windows are constructed of two or more pieces of glass separated by a gas that does not conduct heat well. If you cannot replace older windows, there are several things you can do to make them more energy efficient. First, caulk any cracks around the windows and make sure they seal tightly. Add storm windows or sheets of clear plastic to the outside to create additional air barriers. You can also hang insulated drapes on the inside—in cold weather, open them on sunny days and close them at night. In hot weather, close them during the day to keep out the sun. Windows, doors, and skylights are part of the government-backed ENERGY STAR® program that certifies energy-efficient products. To meet ENERGY STAR® requirements, windows, doors, and skylights must meet requirements tailored for the country's three broad climate region.

Water Heating: Water heating is the third largest energy expense in your home. It typically accounts for about 12 percent of your utility bill. Heated water is used for showers, baths, laundry, dishwashing, and general cleaning. There are four main ways to cut your water heating bills use less hot water, turn down the thermostat on your water heater, insulate your water heater and pipes, and buy a new, more efficient water heater. Other ways to conserve hot water include taking showers instead of baths, taking shorter showers, fixing leaks in faucets and pipes, and using the lowest temperature settings on clothes washers.

There are many strategies that can be used to reduce energy consumption and costs. These strategies might include:

- ✓ Awareness programs to motivate users and occupants to change behaviors that result in saved energy;
- ✓ Operations and maintenance programs, including commissioning to keep equipment running efficiently and to optimize run times;
- ✓ Training programs to educate people on how to reduce energy consumption and costs;

- ✓ Equipment procurement programs to buy the most efficient equipment available (rather than the cheapest first cost) to minimize life-cycle costs, which include energy costs;
- ✓ Commodity procurement programs to buy the least expensive energy source available; and
- ✓ Energy-efficient technology programs to replace existing or conventional equipment with more efficient equipment. This strategy also includes new and emerging energy-efficient technologies.

Review Questions

1. Define energy efficiency and explain with an example why overall efficiency is much less than the individual efficiency?

2. In our daily life we are using incandescent and fluorescent lamps, among these two lamps which one do you prefer and support your answer?

3. It is better to replace incandescent bulbs with fluorescent lamps, and fluorescent lamps with LEDs from energy efficiency point of view. Explain what is energy efficiency and overall efficiency with the help of a supply chain diagram?

4. List out various units in coal fired thermal power plants and characterize the various energy distribution losses in thermal power plant?

5. Mention the conservation steps to enervate the burden of usage of electricity in the world?

6. Write the energy efficiencies of all the resources and characterize their losses. Mention the preventive measures to reduce the losses.

7. What do you understand by energy efficiency and overall efficiency? Explain these with an example of a thermal power plant?

8. Justify compact fluorescent lamps are more efficient than incandescent lamps with an energy flow diagram.

9. Distinguish the energy efficiency of incandescent and fluorescent lamp and discuss percentage conversion losses?

Objective type questions

1. The energy efficiencies of Incandescent and Fluorescent lamps are
   a) Less than LEDs   b) Incandescent lamp > Fluorescent lamp
   c) Fluorescent lamp > Incandescent lamp   d) Both A and C

2. The conversion efficiency of coal fired thermal power plant is
   a) 20%   b) 50%   c) 75%   d) 90%

3. The conversion efficiency of solar energy is
   a) 13%   b) 50%   c) 75%   d) 45%

4. At this time, the most important way to save energy and money in transportation is to

a) Switch to hydrogen-powered cars.   b) Switch to electric engines.

c) Increase the fuel efficiency of motor vehicles   d) Ban cars in cities.

e) Require mandatory mass transportation.

5. The electrical energy produced from a thermal power plant is about------
   a) 1/3rd of primary energy   b) 2/3rd of primary energy

c) Equal to primary energy   d) None of these

6. The efficiency of Incandescent lamp is

   a) Less than 2%   b) 25%   c) 45%   d) 92%

7. Which lamp is better in terms of life span and efficiency?

   a) Incandescent lamp   b) Fluorescent lamp

   c) LED   d) LED and Incandescent lamp

8. BEE stands for

   a) Bharat Electrical and Electronics   b) Board of Electrical Engg

   d) Bureau of Energy Efficiency   b) Bureau of Energy Engg

9. Energy star on any product given by BEE indicates

     a) Most worst material   b) Most efficient material

     c) Highly expensive   d) More energy consumable product

10. Method of reduction of energy in our home is

   a) Replacement of older and lesser efficient equipments with higher efficient materials

   b) Creating awareness of usage of energy

   c) Reduction of heat losses by proper insulation

   d) All of the above

11. The efficiency of a thermal power plant improves with

     a) Increased quantity of coal burnt

     b) Larger quantity of water used

     c) Lower load in the plant

     d) Use of high steam pressures

12. The efficiency of a thermal power plant is

     a) Higher than nuclear energy

b) Depends on calorific value of coal

c) Lesser than solar energy

d) Both a and c

13. For calculating plant energy performance which of the following data is not required

    a) Current year's production

    b) Reference year production

    c) Reference year energy use

    d) Capacity utilization

CHAPTER 7

QUALITY OF LIFE

This chapter deals with the statistics of population in India and how it affects the energy resources with the decrement of quality of life of the people. This effect of energy poverty with lack of basic needs like education, water supply, scarcity of electricity supply, transportation etc., is it because of overpopulation? If it so how to overcome this problem for the encouragement of amelioration of quality of life of rural and urban areas.

7.1 Population Demographics

The demographics of India is the second most populous country in the world with over 121 crores (census 2011) containing 17.5% of world's population. India is projected to be the world's most populous country by 2025, surpassing China, its population reaching

160 crores by 2050. As per population, Population census, Government of India.

Population: 1,210,193,422 (2011 est.)
Growth rate: 1.41% (2009 est.)
Birth rate: 22.22 births/1,000 population (2009 est.)
Death rate: 6.4 deaths/1,000 population (2009 est.)
Life expectancy: 69.89 years (2009 est.)
Male: 67.46 years (2009 est.)
Female: 72.61 years (2009 est.)
Fertility rate: 2.68 children born/woman (2010 est.)
Infant mortality rate: 30.15 deaths/1,000 live births
Population below poverty line: 37% (2010)
GDP per capita: $1,527 (nominal: 133th; 2011)
                    $3,703 (PPP: 129th; 2011)
Unemployment 9.4% (2009 - 10)

Indian population is further classified as,

0-14 years:	31%
15-64 years:	64%
65 and above:	5%

Population below 25 years: 50%
Population below 35 years: 65%

Average Indian is 29 years old (2010 estimate) whereas an average Chinese is 37 years old and an average Japanese is 48 years old. Today India is second most populous country in the world holding 1/6th of the world population and will surpass China by 2030 to become the world's most populous country. The average age of India is 29 years, means that we have a large working force. This sounds good now but soon India as a country has to provide work for a large young generation. This will be followed by lot of energy consumption, prosperity, longevity of population and we reach the Chinese or Japanese average in the near future. The Japanese average age is high (48 years) because they have prospered and are living longer (better health facilities) and hence more old aged people in their society and thus it has become a burden to their society. For every good there is something dangerous that follows. That is how the issue of demographics has to be understood.

The population of India will be as following in the next years.

- 2020:   132 crores
- 2030:   146 crores
- 2040:   157 crores
- 2050:   166 crores

All the statistical data given above is to create awareness, amongst the younger generation, about the need to plan the country's energy needs ahead of times. Demand for energy grows up along with growth in population. The growth in population is always associated with growth in affluence and growth in industrial production which will be demanding more and more energy. Growth in affluence leads to growth in longevity and need for more hospitals. Growth in longevity leads to growth in percentage of aged (non-working) in the society and increase in average age of any given society.

Usage: An example of how European countries were forced to conserve energy during 1980s is cited here to make students understand the seriousness of the issue. Cars have become an important part of middle class families all over Europe and America and each family owned more than one car. With the result all metros are jammed with automobiles. Pollution created through smoke emitted by automobiles was so severe, that it used to disrupt the daily life during winter as visibility was very poor due to smog during day time too. Governments had to frame legislations to control the automobile population within the metros by allowing even numbered cars on certain days of the week and odd numbered cars on the remaining days of the week. That reduced the number of automobiles on the city roads to half. Even today this practice is there in all metros even on Sundays.

India, with large human population has some advantages too. While large population is a disadvantage (as puts pressure on the country's resources), it must be remembered that nearly 65% of the population is below 35 years and hence is a big work force. If India does anything as a society, it affects the whole world (good or bad). Any change, if brought by India, will influence the whole world. India can act from a position of strength and motivate the whole world which otherwise is not possible for small countries.

Poverty reduction and economic growth should be the prime objectives of national policy. Energy tapping is the sign of development. India, with over a billion people, today only produces 660 billion KWh of electricity and over 600 million Indians, a population equal to the combined population of USA and EU, have no access to electricity, and limited access to other clean, modern fuels such as LPG and kerosene. This constrained energy access is reflected in the relatively low Human Development Index of India. Enhancing energy supply and access is therefore a key component of the national development strategy.

Over the past decade, gains in both poverty reduction and economic growth have been significant, and supported by energy growth which has been significantly lower than the economic growth. This reduced energy intensity of the economy, in the period since 2004, has been marked by an economic growth rate of over 9% per annum, which has been achieved with an energy growth of less than 4% per annum.

Some issues:

India's per capita CO_2 emissions are approximately 1 ton per annum, compared to a world average of 4 tons per annum. This is a positive sign, showing that our per capita energy utilization is not dangerously high. In India, the British built the primary railway infrastructure to rapidly transport their raw material to the nearest seaport for outward shipment to Europe to feed the fast growing manufacturing sector. Though it permanently destroyed the traditional time tested water transportation system, an efficient railway network was developed in India. Unfortunately the governments of the independent India have not capitalize on it. Instead they had shifted their priority from low-cost, low-polluting railway transportation to high-cost, high polluting roadways. In 1980, the National Transport Policy Committee noted the cost and energy efficiencies of the rail mode and recommended measures to increase its share in total traffic. However, road-based transportation in India had continued to grow at the expense of rail. The national modal split between rail and road (in percentage terms) in 2000-01 was estimated at 26:74 for freight movement and 18:82 for passenger movement. The share of rail is projected to decline further still.

Problems:

Other problems associated with overpopulation include the increased demand for resources such as fresh water and food, starvation and malnutrition, consumption of natural resources faster than the rate of regeneration (such as fossil fuels), and a decrease in living conditions. However, some believe that waste and over-consumption, especially by wealthy nations, is the force putting strain on the environment, not overpopulation.

- ✓ Inadequate fresh water supply, that it requires extra energy expensive equipments to pure water
- ✓ The overpopulation leads to depletion of natural resources like fossil fuels (petrol, diesel and coal) because of over usage of resources

- ✓ Increased levels of global environmental problems like air pollution, water pollution, noise pollution and soil contamination.
- ✓ Increased in Unhygienic conditions in the surroundings
- ✓ Changes in the climatic conditions like invariability of monsoon seasons and natural disastrous heavy storms, Tsunamis, earthquakes etc.,
- ✓ Leads to deforestation and loss of ecosystems
- ✓ Leads the productive and fertile lands to be futile and arid lands
- ✓ High Infant and child mortality: High rates of infant mortality are caused by poverty
- ✓ Mass species extinction
- ✓ Overpopulation tends to increase in poverty and inflation
- ✓ Low life expectancy

Solutions:

The ratio to examine here is birth rate to death rate. If birth rates exceed death rates, population is increasing. There are two ways to lower the growth rate; decrease births or increase deaths. Because development centralizes around ideas of improving quality of life, and thus life is a central idea to development, advocating a policy of causing death seems contrary to the spirit of the project. So the viable option is lowering birth rates.

In societies like China, the government has put policies in place that regulate the number of children allowed to a couple. Other societies have already begun to implement social marketing strategies in order to educate the public on overpopulation effects. "The intervention can be widespread and done at a low cost. A variety of print materials (flyers, brochures, fact sheets, stickers) needs to be produced and distributed throughout the communities such as at local places of worships, sporting events, local food markets, schools and at car parks (taxis / bus stands)." Such prompts work to introduce the problem so that social norms are easier to implement. Certain government policies are making it easier and more socially acceptable to use contraception and abortion methods.

Limiting birth rates through legal regulations, educating people about family planning, increasing access to birth control and contraception, and extraterrestrial settlement have been suggested as ways to mitigate overpopulation in the future. China and other nations already have regulations limiting the birth rate, with China using the one child policy.

7.2 Energy and Economic Poverty

At the time of independence (1947), India's power generating capacity was 1300 MW, in year 2007) the capacity has reached 138,000 MW (more than 100 times) and at present 2011 the capacity has reached up to 1, 97,000 MW of energy. Yet the country is facing a peak load deficit of 14% and supply deficit of 10%. This scenario is going to continue in the coming decade also and may even worsen. While the food production is satisfactory, electrical power shortages will continue to affect the country's growth. Presently only around 55% of households are electrified (MOSPI 2006) leaving over 20 million households without power. The supply of electricity across India currently lacks both quality and quantity with an extensive shortfall in supply, a poor record for outages, high levels of transmission and distribution (T&D) losses and an overall need for extended and improved infrastructure. Between 2000 and 2008 electricity prices rose by 43.5% and after experiencing a brief drop in prices in 2009, it is now estimated that consumers can expect their utility costs to increase 6.7 to 8 percent annually over the next five years. In addition, gas, heating oil and hydro became subject to the provincial portion of the HST in 2010, further increasing energy costs is due to low-income families live in older houses, with inadequate insulation in attics, walls and basements. While these houses may offer cheaper than-average rent or require lower down payments than more efficient homes, their upkeep is costlier

Unless there is a big breakthrough in the field of energy research, the future generations will have to live with lesser per capita energy. However, nature has given us energy in other forms like solar, wind, tidal and bio matter which are renewable. Students of science and technology have to concentrate more on energy tapping from renewable sources rather than gulping available energy.

In India there are number of factors which led to the situation of energy crisis. There has been sharp rise in the consumption of energy in India since the last decade of 20th century. The year 1991 unlashed the forces of liberalization, privatization sector and accompanying rise in energy, tidal energy etc. These resources have failed to fill the gap of demand and supply of energy due to variety of reasons.

Non-renewable energy is a major source for total energy resource of a country. Coal, oil or gas fired power stations produce electricity. Petroleum derivatives are used in transportation sector. Problem with not renewable energy is that India has to import a major portion of petroleum products as it is not naturally endowed with them in sufficient quantum.

Traditional energy resources like firewood, dried cow dung cake and charcoal are being used in rural India. Such usage of traditional energy resources in inefficient and cause make life miserable for women. To achieve cent percent rural electrification as envisaged by national programmed for rural electrification, availability of ample energy is must.

Hydroelectricity is a cheap source of energy but is inflexible in terms of location. There are many social and environmental concerns such as displacement of tribal's, submergence of forests are associated with hydroelectricity. Other renewable energy resources such as solar power, geo-thermal, tidal power are in nascent stage of development and are commercially unviable.

Nuclear energy is being offered as panacea for energy crisis being faced by India. Nuclear energy is location independent sources of energy i.e. a nuclear reactor can be set in any locality to supply electricity. Further nuclear fuel is cheaper than the petroleum. India has gained a considerable expertise in the development and harnessing of nuclear energy. India's nuclear program is three stage programs which encompasses the use of vast thorium reserves in the county. Nuclear energy is seen to be capable of bridging the gap between the demand and supply of energy in India. Of late, the nuclear energy program faced problem of non availability of natural uranium for rapid expansion of nuclear energy in country. Natural uranium occur in small quantities in India and India cannot import natural uranium from outside as it is not a signatory to NPT and as a consequence NSF refuse to export any nuclear energy related material and technologies to India.

Indo–US unclear deal should be seen in light of the projected benefits of nuclear energy. July 18 deal envisages that US will co operate with India for the development of civilian nuclear technology and use its good offices to ensure NSG rules are modified in a way so that India would be able to receive natural uranium as well as advanced nuclear technologies for civilian use. The deal in turn obliges India to demarcate its civilian as well as military establishment under IAEA regime.

The government of India has been pursuing other avenues also in order to achieve energy security. Government has entered into the agreements with countries like Qatar for the supply of petronet has been established for the purpose of import, storage and then marketing of LNG in India. Government is also actually looking into the feasibility or transnational gas pipelines such pipelines being proposed Iran-Pakistan-India pipeline and Myanmar –Bangladesh India pipeline. It proposes to farm a grid of pipelines and also to rope in china so as to make such venture more secure and financially viable.

ONGC videsh limited is actively investing in overseas petroleum fields by acquiring stakes, farming partnerships with different multinational consortiums to secure hydrocarbons for the country.

India has been given a membership to group of countries involved research in international thermonuclear experimental reactor. This research is aimed at finding means to harness thermo-nuclear energy for peaceful purposes i.e. for the production of electricity. ITER is projected as means that can provide unlimited energy using the principle of nuclear fusion.

Whole India is actively trying to achieve energy security at various international levels. It is imperative that already energy infrastructure should be upgraded and made efficient. National grid system should be able to provide electricity generated out of hydel potential in Himalayas to plains of India. And wind powered electricity from coastal areas to interiors. In other words, country should be capable of transporting energy access regions to energy deficit regions of the country.

There should be greater emphasis on renewable sources of energy. Government should provide funds for research and development of commercially viable methods of harnessing renewable sources of energy. India being a tropical country offers a great potential of solar energy and total energy. Government should subsides solar equipments as it has do for solar crooker. Decentralized HEP and wind power are the commercially viable resources of renewable energy in India today. Small hydroelectric power project made built to satisfy the energy needs of surrounding villages' offers a solution for the electrification of rural areas in mountainous regions. Wind power is already a major source of energy.

State of Tamilnadu, Maharashtra and Gujarat have taken lead in this sector. Government should encourage this private sector imitative by providing the required infrastructure at war footing. So that estimate potential can be harnessed private sector participation should be encourage this private sector imitative by providing the required infrastructure at war footing. So that estimate potential can be harnessed private sector participation should be encouraged.

National policy on electricity now provides that private units can have their own captive power plants and can even sell the surplus electricity to interested buyers. Government needs to provide tax incentives such as rebate in various taxes in order to achieve active private sector participation in energy generation.

Energy crisis in India can treacle by the effective involvement of civil society. In India civil society is inactive in the field of energy conservation unlike the western countries. Energy conservation is the key civil society can lead the front by educating masses about the need to conserve energy. Energy conserved is energy earned.

Generation of electricity is associated with generation of affluent gases, fly ash and other pollutants. The more the energy tapped or generated, the less will be left for future. One should not forget that energy available is a constant and we can only change the form for our purposes.

Key Findings

- Energy poverty is expected to rise without intervention as the result of rising energy costs which are expected to increase 6.7 to 8 percent annually over the next five years.
- Energy poverty directly and indirectly impacts resident's health and can result in disconnection and eviction leading to homelessness.
- Energy poverty can be eradicated by increasing income, regulating energy pricing and improving energy efficiency of homes.
- Municipal, Provincial and Federal governments all have an integral role to play in eliminating energy poverty.

Solutions of energy poverty:

To eliminate the energy poverty, it has to access to affordable and energy efficient homes, increasing social assistance rates, and providing a living wage to the working poor. In the absence of such initiatives, the research proposes three possible ways to move vulnerable households out of energy poverty:

1. Increase income
2. Regulate energy pricing
3. Reduce home energy usage
4. Energy Conservation

1. Increase Income

There are several local programs that provide financial assistance to households experiencing energy poverty.

Winter Warmth provides assistance for natural gas bills through Union Gas or Enbridge Gas. It is designed for low-income families and individuals living at/or below the poverty line who have exhausted all other sources of financial support.

Low Income Energy Assistance Program (LEAP) provides emergency financial assistance from local electricity providers to low-income energy customers through community organizations.

2. Regulate Energy Pricing:

Continuous monitoring of meters that can provide stable and predictable electricity pricing and to provide consumers with an incentive to shift some of their consumption away from periods of high total consumption to periods of low demand and save money on their bill. Smart metering, which tracks how much electricity a household uses and when, poses particular challenges for low-income households who may not have the same flexibility to alter their energy use during the day as someone who is at work and out of the home. The actual impact of time-of-use pricing depends on the precise rate structure and on the extent to which a household shifts its electricity use away from periods when the 'onpeak' rate applies.

Time-of-use pricing and smart metering also creates challenges for social housing providers. Since most social housing tenants do not pay directly for electricity they are not exposed to higher peak prices and have no corresponding incentive to reduce peak use. Thus, time-of-use pricing could result in increased electricity bills that housing providers will have to meet

3. Reduce Home Energy Usage

Home Energy Efficiency Programs (HEEPs) establish policies for energy efficiency and conservation through setting standards, providing support and incentives for retrofits, and encouraging the use of more efficient furnaces and other appliances. However, low-income households are unlikely to have enough disposable income to take advantage of such opportunities. Reports also note that costs for utility incentive programs are recovered through utility rates paid by all ratepayers, despite low-income ratepayers' inability to participate in the programs. This creates the perverse situation whereby low-income ratepayers subsidize the energy efficiency upgrades of their higher-income counterparts Landlord-tenant relationships can also generate barriers to landlord participation in HEEPs. Where tenants pay energy bills, landlords lack the incentive to pay money to curb those costs; where landlords pay energy bills, tenants have no incentive to conserve, which may undermine retrofit investments

4. Energy Conservation:

It is high time that every citizen must start thinking about conservation of energy in every act in everyday life. For example, energy can be conserved in our daily lives as depicted below.

- Use water sparingly for bathing and flushing
- Eat at home as far as possible
- Use mass transport as far as possible
- Remember S.O.S. (switch off something) while leaving the room
- Use books instead of surfing net too often for reading
- Avoid hot water bath in tropical areas
- Avoid overloading refrigerators
- Reduce utilization of air conditioners (use windows cleverly)
- Use bicycle where ever possible
- Use less plastics (carry your own bag to the market)
- Grow a tree (let your waste water go to tree)

Economic Poverty

Even more than 60 years after independence from almost two centuries of British rule, large scale poverty remains the most shameful blot on the face of India. Of its nearly 1 billion inhabitants, an estimated 350-400 million are below the poverty line, 75 per cent of them in the rural areas. More than 40 per cent of the population is illiterate, with women, tribal and scheduled castes particularly affected. According to a 2005 World Bank estimate, 26.1% of the total Indian population falls below the international poverty line of US$ 1.25 a day (PPP, in nominal terms₹ 21.6 a day in urban areas and ₹ 14.3 in rural areas). The growth of the middle class indicates that economic prosperity has indeed been very impressive in India, but the distribution of wealth has been very uneven. The main causes of poverty are illiteracy, a population growth rate by far exceeding the economic growth rate for the better part of the past 50 years, protectionist policies pursued since 1947 to 1991 which prevented large amounts of foreign investment in the country.

Eradication of Poverty:

The most important initiative has been the supply of basic commodities, particularly food at controlled prices, available throughout the country as poor spend about 80 percent of their income on food. Eradication of poverty can only be a very long-term goal in India. Poverty alleviation is expected to make better progress in the next 50 years than in the past, as a trickle-down effect of the growing middle class. Increasing stress on education, reservation of seats in government jobs and the increasing empowerment of women and the economically weaker sections of society, are also expected to contribute to the alleviation of poverty.

7.3 Quality of Life in Rural and Urban areas

The term quality of life is used to evaluate the general well-being of individuals and societies. The term is used in a wide range of contexts, including the fields of international development, healthcare, and politics. Quality of life should not be confused with the concept of standard of living, which is based primarily on income. Instead, standard indicators of the quality of life include not only wealth and employment, but also the built environment, physical and mental health, education, recreation and leisure time, and social belonging.

Since independence, India has paved the way through democracy for social development. India has been implementing national strategies and plans through Five Year Plans (F.Y.P.), various multi faceted development schemes and programs. These programs, backed by large human and financial resources, have been successful in achieving the predetermined goals in the areas of sustained economic growth, education, health, sanitation, housing and employment, as well as other related fields, so that minimum needs are duly taken care of and a decent standard of life attained.

Eradication of poverty and provision of basic minimum services to all citizens are integral elements of any strategy to improve the quality of life. No developmental process can be sustainable unless it leads to visible and widespread improvement in these areas. India believes that poverty anywhere is a threat to prosperity everywhere and that concerted international action is essential to ensure global prosperity and better standards of life for all. Based on this belief, India has actively played a positive, constructive role, inter alia, in the deliberations of the UN, its specialized agencies and various intergovernmental mechanisms.

The Eighth Plan (1992-1997) had identified "human development" as its main focus. During this plan period, the indicators of social development have shown a significant improvement. 1995-96 witnessed a very satisfactory growth rate in GDP of 7.1 per cent. The momentum of growth has been maintained in 1996-1997, thus providing increasing evidence that the growth potential has improved as a result of the processes of deregulation and globalization initiated by the government.

The Ninth plan (1997-2002) was launched in the 50th year of India's Independence. The objectives of the Ninth Plan arose from the Common Minimum Program of the Government; a few pertinent of these objectives are as follows:

1. Economic growth and overall development;
2. Human development with emphasis on health, education and minimum needs, including
   protection of human rights and raising the social status of the weak and the poor; and
3. Directly targeted programs for poverty alleviation through employment generation, training and building up asset endowment of the poor.

Perhaps the most commonly used international measure of development is the Human Development Index (HDI), which combines measures of life expectancy, education, and standard of living, in an attempt to quantify the options available to individuals within a given society. The HDI for India is 0.619, which gives the country a rank of 128th out of 177 countries. The three parameters used for calculating the HDI are identified. As such these parameters combine Life expectancy at birth (63.7%), adult literacy rate (64.15) and GDP per capita 93452), have been used to calculate the GDP value.

One of the keys to making important changes is recognizing that what *you* want and need is not necessarily the same as what other people think is best for you or what they would want and need for themselves - 'Quality of Life' is subjective.

Objective of QOL:

The objective quality of life measures are built on the basis of hard variables i.e., the data from the municipal or governmental institutions and organizations which may include financial accounts, civil state records, medical statistics, pollution levels, and other pieces of factual information gathered by the institutions routinely. This approach aims at investigating the society as a whole by looking, in the most general sense, at the set of macroeconomic, social, demographic indicators which determine the conditions of life and the way people live. As an objective measure, the quality of life may be defined as an interrelation of the four determinants of the vital functioning and activity of the population.

For Example, the quality of population is inferred from the population demographic structure, the reproductive process and marital behavior, the physical, psychic, and moral health of the population as well as their educational level and proficiency. The material welfare is determined by the standards of living, income differentiation, housing, telecommunication, trade, education, culture, health system, mass media capacities. More simply, parameters such as access to basic services (drinking water, sanitation etc.) and facilities (market, park, bank etc) are needed to depict QOL of a community. The social system quality relies upon the state and/or private provision in cases of the (permanent or temporary) disablement, citizens' rights for education, employment, recreation, private property and personal protection, political system stability, individual's inclusion to the social infrastructure, race and sex equality, social stability. The ecology state is influenced by the state of air and water sources (surface and ground), the level of chemical, radioactive, heavy metal pollution, etc. The list is far from being complete, and some items may be related to more than one category.

So, each of the mentioned properties and measures, being expressed via a system of statistic indicators, should then be integrated into a measure of the overall quality of life. There can be no doubt however, that QOL of a community or a social group or a nation for that matter is determined by the common members or citizens who are largest in number. Thus a balanced combination of the major aspects contributing to a QOL statement would be required. Further, though the set of parameters and their respective benchmarks may vary according to the inherent character of the community, for comparability across communities, a global set would be useful.

Characteristics of Quality of Life:

Lower Income Countries or Least Developed Countries are characterized by a marginal physical environment. The countries that fall into this category are, most of African countries, and many of the Asian countries. The LDC countries have the following characteristics:

1. Little to no build-up of agriculture

2. Low energy production and consumption

3. Subsistence farming (farm only for themselves)

4. Large percentage of the population is under 15

5. Infant mortality rate is high

6. Poorly developed trade and transportation

7. Poor medical facilities

8. High illiteracy and unemployment rates

9. GNP is under $3,000 per year.

Improvement of Quality of life:

1. The improvement of the quality of life of human beings is the first and most important objective of every human settlement policy. These policies must facilitate the rapid and continuous improvement in the quality of life of all people, beginning with the satisfaction of the basic needs of food, shelter, clean water, employment, health, education, training, social security without any discrimination as to race, color, sex, language, religion, ideology, national or social origin or other cause, in a frame of freedom, dignity and social justice.

2. In striving to achieve this objective, priority must be given to the needs of the most disadvantaged people.

3. Economic development should lead to the satisfaction of human needs and is a necessary means towards achieving a better quality of life, provided that it contributes to a more equitable distribution of its benefits among people and nations. In this context particular attention should be paid to the accelerated transition in developing countries from primary development to secondary development activities, and particularly to industrial development.

4. Human dignity and the exercise of free choice consistent with over-all public welfare are basic rights which must be assured in every society.

5. The nations must avoid the pollution of the biosphere and the oceans and should join in the effort to end irrational exploitation of all environmental resources, whether non-renewable or renewable in the long term. The environment is the common heritage of mankind and its protection is the responsibility of the whole international can unity. All acts by nations and people should therefore be inspired by a deep respect for the protection of the environmental resources upon which life itself depends.

6. All countries, particularly developing countries, must create conditions which make possible the full integration of women and youth in political, economic and social activities, particularly in the planning and implementation of human settlement proposals and in all the associated activities, on the basis of equal rights, in order to achieve an efficient and full utilization of available human resources, bearing in mind that women constitute half of the world population.

The goals for the development of Quality of life

Goal 1: Eradicate extreme poverty and hunger

Target 1- Halve, between 1990 and 2015, the proportion of people whose income is less than 50 Rs. a day.

Target 2 - Halve, between 1990 and 2015, the proportion of people who suffer from hunger.

Goal 2: Achieve universal primary education

Target 3 - Ensure that, by 2015, children everywhere, boys and girls alike, will be able to complete a full course of primary schooling.

Goal 3: Promote gender equality and empower women

Target 4 - Eliminate gender disparity in primary and secondary education, preferably by 2005, and to all levels of education no later than 2015.

Goal 4: Reduce child mortality

Target 5 - Reduce by two thirds, between 1990 and 2015, the under-five mortality rate

Goal 5: Improve maternal health

Target 6 - Reduce by three quarters, between 1990 and 2015, the maternal mortality ratio.

Goal 6: Combat HIV/AIDS, malaria and other diseases

Target 7 - Have halted by 2015 and begun to reverse the spread of HIV/AIDS

Target 8 - Have halted by 2015 and begun to reverse the incidence of malaria and other major diseases.

Goal 7: Ensure environmental sustainability

Target 9 - Integrate the principles of sustainable development into country policies and programs and reverse the losses of environmental resources.

Target 10 - Halve by 2015 the proportion of people without sustainable access to safe drinking water.

Target 11 - By 2020 to have achieved a significant improvement in the lives of at least 100 million slum dwellers.

Goal 8: Develop a Global Partnership for Development

Target 12- Develop further an open, rule based, predictable, non-discriminatory trading and financial system

Target 13- Address the special needs of the least developed countries

Target 14- Address the special needs of landlocked countries and Small Island developing States.

Target 15- Deal comprehensively with the debt problems of developing countries through national and international measures in order to make debt sustainable in the long term

Key Findings

1. Central governments play a critical role in determining the prosperity and growth of cities

2. Balanced urban and regional development can be achieved through consistent and targeted investments in transport and communications infrastructure.

3. Cities are becoming more unequal

4. High levels of urban inequality are socially destabilizing and economically unsustainable

5. Focused and targeted investments and interventions can significantly improve the lives of slum dwellers

6. Cities provide an opportunity to mitigate or even reverse the impact of global climate change as they provide the economies of scale that reduce per capita costs and demand for resources.

7. Evidence shows that compact and well-regulated cities with environmentally-friendly public transport systems have a positive environmental impact

8. Sea level rise could have a devastating impact on coastal cities

9. Commitment to pro-poor, inclusive urban development

10. Coordination and collaboration between national, provincial and local authorities can achieve harmonious regional and urban development.

Source: The Economist Intelligence Unit's Quality-of-Life Index" (PDF). The Economist;

http://www.economist.com/media/pdf/QUALITY_OF_LIFE.pdf.
Retrieved on 2009-10-03)

Review Questions

1. Compare the quality of life in rural and urban areas?

2. What do you understand by the term quality of life and how a nation's quality of life is determined?

3. What are the most important needs to be satiable for individual, family and a society for better quality of life?

4. What is energy poverty and what are the different causes of poverty in different?

5. What do you understand the term demographics and discuss some issues related to population demographics in India?

6. Even a vast development of technology in this 21st century, but 100% electrification is not provided in some rural and urban areas in India. As a citizen of India, give your suggestions for providing 100% electrification in rural and urban areas?

7. Even more than 60 years after independence from British rule, large scale poverty remains on face of India. Mention the reasons why India still in poverty and give your suggestive measures to alleviate poverty?

8. Examine the sources of electrification in rural areas and urban areas?

9. Discuss the problems and solutions of over population in India.

10. What is quality of life and how can we evaluate QOL?

11. Let us consider you are having full powers in your village then how can you ameliorate the quality of life of your village?

12. What is quality of life? What are the factors considered to estimate the quality of life of a nation? How energy is related to quality of life and discuss the quality of life in rural and urban areas.

13. Illustrate your ideas of amelioration of quality of life of in Rural and Urban areas?

14. Examine the sources of electrification in rural areas and urban areas?

15. How energy is related to poverty? Discuss poverty in India and methods to alleviate poverty.

16. Discuss the reasons of not providing 100% electrification in some rural and urban areas in India and what are the necessary steps had to take into consider in order to provide 100% electricity?

17. What is energy poverty and what are the different causes of poverty in different?

Objective type Questions

1. Energy poverty in rural areas is high compared to urban areas because

   a) Non-availabilty of Transportation
   b) Lack of awareness among people i.e low literacy rate
   c) Lack of affordability   d) All of the above

2. The average age of Indian according to the 2010 population demographics is

   a) 41 years   b) 29 years
   c) 47 years   d) 54 years

3. Quality of life is considered high

a) When we consume more energy
b) When we consume less energy
c) When energy is available just sufficient to meet our basic needs
d) When we able to produce energy by unconventional methods

4. The approximate income level per day per person fixed in India to be rated as below poverty

   a) Rs 25   b) Rs 75   c) Rs 100   d) Rs 200

5. As populations in India grow, so does the need for wood to build houses, stores, and other buildings. Select the statement that best describes the long–term global effects of this behavior.

   a. People will need to plant more trees and in a few months the entire problem will be taken care of.

b. If we control the amount of forests cut, the people will go somewhere else to find wood. This is not a concern as long as they are not cutting down the forests in North America.

c. Populations are falling in the rest of the world so the extra wood can come from other countries that have a surplus.

d. In order to meet the demand, forests have been cut down at an alarming rate. If we do not find another source of building materials, our forests, an important natural resource, will all disappear.

6. The percentage of India's population in world is

    a) 18%   b) 5%   c) 10%   d) 12%

7. The major percentage of India's population is in between

    a) 0-14 years   b) 15-64 years
    c) above 65 years   d) None of the above

8. The method of decreasing population is

    a) Decreasing Birth/Death ratio
    b) Increasing Birth/Death ratio
    c) Controlling birth rate by doing family planning
    d) Both a and c

History is the teacher of life, and life is sustained by energy. A critical examination of the periods in the history in which major changes in energy use have occurred can be helpful to understand better our present relationship with energy as well as to guide our decisions and choices. The human recognition of energy systems starting from fire and subsequent changes in the energy systems by each decade have been explained in this chapter by considering into three sections as Energy systems in Past, Present and Future. And this chapter also dealt with the technologies of more sustainable energy sources in lieu of technologies with environmental issues.

8.1 Energy Systems in the Past

Energy Systems: An energy system may be thought of as an interrelated network of energy sources and stores of energy, connected by transmission and distribution of that energy to where it is needed. The transformation from stores of energy in food to work and subsequent dissipation of energy is an example of such a system. The starting point of all energy in this "food chain" or "energy chain" (considering only the vegetable and cereal part of our food) is the sun.

- In the beginning, the sun was the only source of energy. People would gather food and hunt during the day, and when the sun went down at night, they would search for shelter from the cold.

- Eventually someone noticed that the fires that were sometimes started by the lightning, brought comfort and warmth. Fire is one of the most important discoveries of all times. With fire, people could cook their food, and warm themselves on cold nights.
- Fire is essentially converting energy into heat by releasing stored energy. Wood was the main source of this energy for thousands of years. The use of wood for fuel grew and grew, especially after Europeans discovered uses for iron and steel. As more and more wood was used in the steel making furnaces, Europe began to experience rapid deforestation, especially from the 15th to the 18th century. It was evident that the use of wood as fuel could not continue at its current rate.
- By the middle of the 18th century, much of Europe was experiencing an energy crisis due to lack of timber. As a result, coal became the major source of fuel. Coal was a plentiful source of energy, and England led the change into this new era that we call the Industrial Revolution.
- Coal is a very abundant source of energy. At the current rate of use, coal will last another 200 years. Coal represents about 78% of the world's available fossil fuels, [Moyers 2002] and can be found in many countries around the world. China and the United States together share 50% of the world's coal, but for now there is enough dispersed in different locations to make it a plentiful source of energy.
- Coal is used in many industrial applications, especially in areas such as the steel industry. Another one of the main uses of coal is in the generation of electricity. 38% of the world's electricity is generated by coal [Moyers 2002]. Even though coal is a very abundant energy supply there are concerns about its role in pollution. Pollutants released by burning coal include sulfur, carbon dioxide, and mercury.
- Oil has been used as a source of energy for thousands of years. Sumerians, Assyrians and Babylonians used crude oil that seeped

out of the ground along the Euphrates river [energy quest] for lighting and medicine.

- Almost all plastic, and the fertilizers that we use to help our crops grow come from petroleum. Oil is the main source of fuel for transportation vehicles, and the roads that we drive our cars on come from oil.

- The quest for more powerful energy sources was propelled by the inventions and discoveries of the Industrial Revolution. The need for large quantities of accessible, dependable, and transportable energy encouraged the exploration of energy sources. The inventions of the Industrial Revolution provided the equipment to further mine or drill the already visible deposits of coal and oil.

- Steam power was developed in the 1600's in conjunction with coal mining to help pump water out of the mines. It had been known since ancient times that heat could be used to produce steam, which could then do mechanical work. However, it was only in the late eighteenth century that commercially successful steam engines were invented. The first commercially successful steam engine was invented by Thomas Savery (1650-1715), an English military engineer.

- The sun was also studied as an energy source in the 18th century. In 1767, the first solar thermal collector was developed.

- In 1839, Alexandre Becquerel discovered that an electric current could be generated when certain elements were exposed to light. The scientific explanation of this phenomenon by Albert Einstein, called photoelectricity (light-induced electricity), came much later in 1905. Photoelectricity is the basis of the photovoltaic cells, now used to convert light into electricity.

- In 1820, the advances in mechanical and materials engineering made the railroad the most efficient and fastest means of transportation. Coal and wood were used as the primary fuel source for the steam engine.

- As early as 1816, natural gas was piped into cities for domestic uses such as cooking, home illumination, and street lighting. The steam engine was used to pump water into homes and sewage away from homes.
- In 1859, when petroleum was drilled in Titusville, Pennsylvania, an apparently plentiful energy source began to replace coal. Oil was distilled into kerosene (referred to as coal oil) and used as a lamp oil. It replaced dwindling supplies of whale oil used for lamps. There were many reasons oil became a more desirable fuel source than coal: it was easy to obtain and transport; it emitted less particulate pollution than coal; it replaced scarce whale oil as a fuel for lamps; and coal had become an unreliable fuel source because of the labor issues surrounding the mining of coal.
- But the most significant use of crude oil was as the liquid fuel for the internal combustion engine, designed in 1861 by Nikolaus August Otto. The internal combustion engine became one of the most influential inventions of the Industrial Revolution.
- In 1879, Thomas Edison invented the incandescent light bulb -- a major step in the human use of storable energy leading eventually to large-scale electrification.
- In the late 19th and early 20th centuries the steam turbine, using coal as a fuel, was developed as a cheap power source that generated electricity.

8.2 Energy Systems in the Present

- In 1910, Henry Ford opened the 60-acre Highland Park automotive plant with a moving assembly line. This was the beginning of an eventually enormous use for fossil fuels. Fossil fuels were used not only to propel the automobiles that were made at the plant, but also to generate electric power for the automotive plant.

- Energy technologies developed rapidly during the 20th century. Although the current version for solar thermal collectors was designed in 1908, they were not developed well enough for mass distribution. In the 1920's, 30's, and 40's, there was large-scale construction and development of hydroelectric plants/dams to support increasing population in the Southwest.

- In 1938, Otto Hahn and Fritz Strassman demonstrated nuclear fission and within four years (1942), Oak Ridge, Tennessee, was chosen as the site for the first functional nuclear reactor plant, and for the preparation of uranium and plutonium used to the create the atomic bomb at Los Alamos.

- The first nuclear chain reactor was demonstrated at the University of Chicago in December 1942. In July 1945, the testing of the first atomic bomb at Alamogordo, New Mexico, demonstrated the technology used to release nuclear energy on a large scale. In 1957, the first commercial nuclear power plant opened in Shippingport, Pennsylvania.

- The first large scale use of photovoltaic (PV) solar energy in conjunction with satellite technology developed in the 1950's. The United States Vanguard I was the first PV-powered satellite.

- By the early part of the 20th century, crude oil and its products had become an indispensable part of the industrial economy. James Young had patented a process in England in 1850 to distill oil from coal and shale. Oil refining is not just about gasoline. The distilled chemicals from crude oil have many purposes -- for example, petroleum is used for plastics manufacturing. Young's process of fractal distillation forms the basis of the world's oil refining industry.

- India ranks sixth in the world in total energy consumption and needs to accelerate the development of the sector to meet its growth aspirations. The country, though rich in coal and abundantly endowed with renewable energy in the form of solar, wind, hydro and bio-energy has very small hydrocarbon reserves (0.4% of the world's reserve).

- India, like many other developing countries, is a net importer of energy, more than 25 percent of primary energy needs being met through imports mainly in the form of crude oil and natural gas. The rising oil import bill has been the focus of serious concerns due to the pressure it has placed on scarce foreign exchange resources and is also largely responsible for energy supply shortages. The sub-optimal consumption of commercial energy adversely affects the productive sectors, which in turn hampers economic growth.

- If we look at the pattern of energy production, coal and oil account for 54 percent and 34 percent respectively with natural gas, hydro and nuclear contributing to the balance. In the power generation front, nearly 62 percent of power generation is from coal fired thermal power plants and 70 percent of the coal produced every year in India has been used for thermal generation.

- The distribution of primary commercial energy resources in India is quite skewed. 70 percent of the total hydro potential is located in the Northern and Northeastern regions, whereas the Eastern region accounts for nearly 70 percent of the total coal reserves in the country. The Southern region, which has only 6 percent of the total coal reserves and 10 percent of the total hydro potential, has most of the lignite deposits occurring in the country. On the consumption front, the industrial sector in India is a major energy user accounting for about 52 percent of commercial energy consumption.

8.3 Energy Systems for the Future

- **Wind:** Wind energy is a mature option in sustainable energy with great potential and a rapid development over the past 25 years. In 2007 the installed capacity in Denmark was about 3 GW and wind turbines produced electricity equal to 20%of the total Danish electricity demand. In 2008 the global installed wind power capacity was about 100 GW [2, 3]. For some years, world wind capacity has been doubled every three to four years. In the years ahead the growth rate is expected to be higher in the USA and Asia. Despite this technological development, and rapid growth in a few countries, wind today provides only a small percentage of the world's electricity.

- **Photovoltaic's:** Photovoltaic (PV) devices, otherwise known as solar cells, convert light directly into electricity. PV technology is modular and contains no moving parts. Solar cells were the fastest-growing renewable energy technology market in 2005, with a global annual growth rate of more than 40%, and this trend continued in 2006. Growth has been dominated by grid-connected distributed systems in Germany and Japan. The status of PV technology, its potential and R&D challenges were addressed comprehensively by the EU-supported publication A Vision for Photovoltaic Technology compiled by the Photovoltaic Technology Research Advisory Council. These R&D challenges are presently being analyzed more detail in a study called the PV Strategic Research Agenda (SRA), which was published in 2007.

- **Solar Thermal:** Solar thermal heating is a long-established technology for space heating and domestic hot water supply. New applications are emerging for industrial processes, where solar energy could replace fossil fuels or electricity. For solar thermal devices the average annual market growth rate has been 17–20% in recent years. The most dynamic market areas are China and Europe. In absolute terms the European solar thermal market is dominated by Germany (~50%), followed by Greece and Austria (~12% each). Europe's present solar thermal capacity provides around 0.15%of the overall EU requirements for hot water and space heating. In general, costs per unit area decrease with the size of the system. Solar thermal systems connected to a district heating network are therefore more cost-effective than systems for single family houses. Solar thermal systems traditionally include short-term hot water storage capacity in the range50–75 l per m^2 of collector. Seasonal storage of around 2,000l per m^2 has been investigated, but is still considered to be at the R&D stage. A relatively new market for solar thermal units is industrial process heat. Low-temperature process heat, in the range achievable by traditional solar collectors, is needed in many industries.

- **Biomass based fuels for transport:** There are several motivations to provide alternative transport fuels based on biomass as a raw material. It will be transport fuel with low CO_2 emissions, it will reduce the dependence on imported fossil fuels in the Western world and it is possible to further develop a domestic industry based on liquid fuels. Liquid transport fuels based on biomass can be produced by several different means such as biodiesel from rape, ethanol by fermentation and by the GTL-technology (Gas-To-Liquid). The GTL-technology has the potential to obtain a high biomass to liquid conversion efficiency, and it should be possible to develop the technology so that a broad range of solid input fuels can be applied. A disadvantage is that GTL-plants are relatively large and complicated.

- **Thermal fuel conversion – combustion, gasification and pyrolysis of biomass:** The thermal conversion of biomass and waste into power, heat and process energy is today the world's largest contributor of CO_2 neutral energy and will also in the future provide a large share of CO_2 neutral energy supplies. A very broad range of thermal based technologies are used today, and some emerging thermal technologies will also be used in the future. A range of research challenges persists including: Increased biomass fuel share in power plant boilers, increased electrical efficiency of waste and biomass combustion plants and reduced operational problems, development of mature and flexible pressurized gasification technologies and development of reliable biomass pyrolysis reactors.

- **Gasification:** Industrial-scale gasification is currently mostly used to produce electricity from fossil fuels such as coal, where the syngas is burned in a gas turbine. Four types of gasifier are currently available for commercial use: counter-current fixed bed, co-current fixed bed, fluidized bed and entrained flow. Gasification is also used industrially in the production of electricity, ammonia and liquid fuels (oil) via Integrated Gasification Combined Cycles (IGCC), with the possibility of producing CH_4 and H_2 for fuel cells. IGCC is also a more efficient method of CO_2 capture as compared to conventional technologies. IGCC demonstration plants have been operating since the early 1970s and some of the plants constructed in the 1990s, are now entering commercial service. In the early research stage is microbes for the in-situ-coalmining producing methane as a product of digestion.

- **Nuclear Energy:** Nuclear fission energy is the major CO_2 emission free source; it provides 15% of the world electricity production and 7% of the total energy consumption. Globally, 440 reactors are in operation in 31 countries with most of the nuclear generation capacity being in Europe, the US, and Southeast Asia. Due to the high capital cost of nuclear reactors and low fuel prices nuclear energy is used predominantly for base load electricity production. In Europe, nuclear accounts for 20% of the generation capacity but provides 31% of the electricity generation. The technology is fully developed and available to the market. However, the majority of existing nuclear power units was built in the 1970s and 1980s. After 1990, nuclear power globally faced stagnation. Construction of nuclear power plants, however, continued in the Far East, especially in Japan and South Korea. Since 1990 the global installed capacity has increased only slightly to the present value of 370GWe.Nuclear power is not vulnerable to even high fuel price fluctuations, and as it is based on uranium sources that are widely distributed around the globe, fuel supply is not strongly affected by geopolitical issues. In addition, because many years' worth of nuclear fuel can be stored in a small area, the presence of local uranium resources is not a pre-condition for nuclear energy security.

- **Fusion Energy:** A fusion reactor would "burn" the isotopes deuterium and tritium at moderate pressure and at a temperature of 150million Kelvin. A fusion reactor will produce much less radioactive waste than a fission reactor. Fusion plants are inherently safe as the reactor only contains enough fuel to feed the fusion processes for the next few seconds. The main cost of fusion energy will be in constructing the power plant, while the cost of fuel is negligible. Fusion power will therefore be most economical when run as base load, though it can easily contribute to a sustainable energy mix.
- **Geothermal Energy:** Geothermal energy is heat from within the earth. The steam and hot water produced inside the earth can be used to heat buildings or generate electricity. Geothermal energy is a renewable energy source because the water is replenished

by rainfall and the heat is continuously produced inside the earth.

The main uses of geothermal energy are:

1. Direct use and district heating systems which use hot water from springs or reservoirs near the surface.
2. Electricity generation in power plant requires steam at very high temperature (300 to 700 degrees Fahrenheit). Geothermal power plants are generally built where geothermal reservoirs are located within a mile or two of the surface.
3. Geothermal heat pumps use stable ground or water temperatures near the earth's surface (less than 100 meters) for space heating.

- **Hydro, Ocean, Wave and Tidal:** This group of energy supply technologies is based on the use of potential, kinetic or thermal energy of water as energy source and show different stage of development. Hydropower and pumped hydro storage systems have for many years been fully commercially competitive in many parts of the world. On the other hand ocean energy, including wave and tidal are at an early stage of development.

Large hydro remains one of the lowest-cost generating technologies, although environmental constraints, resettlement impacts and the limited availability of sites have restricted further growth in many countries. Large hydro supplied 16%of global electricity in 2004, down from 19% a decade ago. Large hydro capacity totaled about 720 GW worldwide in2004 and has grown historically at slightly more than 2% annually. China installed nearly 8 GW of large hydro in 2004, taking the country to number one in terms of installed capacity (74 GW). With the completion of the Three Gorges Dam, China will add some 18.2 GW of hydro capacity in 2009.

Small hydropower has developed for more than a century, and total installed capacity worldwide is now 61 GW. More than half of this is in China, where an ongoing boom in small hydro construction added nearly 4 GW of capacity in2004.

Ocean currents, some of which run close to European coasts, carry a lot of kinetic energy. Part of this energy can be captured by submarine "windmills" and converted into electricity. These are more compact than the wind turbines used on land, simply because water is much denser than air. The available power is about 1.2 kW/m² for a current speed of 2 m/s, and 4 kW/m² for a current of 3 m/s. The main European countries with useful current power potential are France and the UK.

Ocean tides can be exploited for only four or five hours per cycle, so power from a single plant is intermittent. A suitably designed tidal plant can, however, operate as a pumped storage system, using electricity during periods of low demand to store energy that can be recovered later. The only large, modern example of a tidal power plant is the 240 MW LaRance plant, built in France in the 1960s, which represents91% of world tidal power capacity. Wave energy can be seen as stored wind energy, and could therefore form an interesting partnership with wind energy. Waves normally persist for six to eight hours after the wind drops, potentially allowing wave power to smooth out some of the variability inherent in wind power.

- **Fuel cells:** Fuel cells are at the point of breakthrough as a most versatile and efficient energy conversion technology. They have strong links with renewable technologies, such as wind, solar and wave power, and they will be central to any future "hydrogen society", with its promise of a release from dependence on fossil fuels. Denmark is playing a significant role in the development of fuel cells, all the way from fundamental research to consumer applications. Low-temperature fuel cells, notably PEMFCs could replace car engines and are already being used in commercial uninterruptible power supplies, such as those made by the Danish company Dantherm.

High-temperature fuel cells (solid oxide fuel cells (SOFCs) and molten carbonate fuel cells (MCFCs) are fuel-flexible, highly efficient and environmentally clean. They can run on fuels such as natural gas, biogas and methanol.

The application areas for fuel cells fall into three main markets: stationary, transport, and portable. The stationary market ranges from small (≤ 1 - 5 kW) CHP units for single households to 100-1000 kW CHP units for district heating; and multi-MW units for power generation. Fuel cells may become important in the transportation sector in hybrid cars, buses, trucks and trains. The first commercial fuel cells are now appearing in portable applications and backup power systems.

- **Hydrogen:** Hydrogen is an energy carrier, not an energy source. Realizing hydrogen as an energy carrier depends on low-cost, high-efficiency methods for production, transport and storage. Hydrogen can be produced by many technologies, based on fossil and sustainable fuels. Thermal and thermo chemical processes use heat to release hydrogen and are the most mature technologies. Electrolytic processes use electricity to produce hydrogen. Here renewable sources such as wind can be considered. An electrolyser is based on the same principles as a fuel cell, but the process is reversed, i.e. electricity is used. Electrolysis will likely play an important role in any future non-fossil energy scenario, not only in the hydrogen society. Current costs of electrolysers are high but declining. The degree of sustainability of the hydrogen production strongly depends on the feedstock used. Ultimately, hydrogen fuel could be produced in association with CCS leading to low-emission transport fuels. Photolytic processes offer a challenging, long-term potential for a sustainable hydrogen production and have to be further developed.

8.4 Sustainable Development Issues

Sustainable development is development that meets the needs of the present without compromising the ability of future generations to meet their needs. Sustainable development has three aspects:

1. Social (people)
2. Environmental (planet)
3. Economic (profits)/prosperity

All development affects all three aspects. All three aspects are interdependent. Thus, being mindful of these interdependencies in management and leadership decisions will result in the best overall solution is to maximize success and minimizes any negative social, environmental, and economic costs.

The environmental (planet) aspect is significantly affected by energy consumption and management, including: the entire national power infrastructure and distribution, transportation, plus the construction and renovation of all residential, commercial, and industrial facilities.

Environmental Sustainability:

During the last century, while fossil fuels were abundant and cheap, those fuels filled a majority of our energy conversion needs. The mounting problem is that combustion emissions have fouled the environment in a number of ways, resulting in increases in respiratory illnesses, mercury pollution, and arise in global temperatures. The quantity of easily retrieved fossil fuels is significantly depleted. Coal is still relatively abundant, but it does not burn cleanly. Technologies need to be developed to both mine the coal safely and to burn it cleanly. It is also apparent that there is a need to protect and allow the environment to regenerate. The effect of using fossil fuels extensively and inefficiently is that we are simultaneously poisoning the environment and ourselves. Whole world's of the government and researchers are taking action to implement clean energy technologies. In response to clear signs of increased cost from continuing to use fossil fuels and scientific evidence showing that by burning fossil fuels we are initiating a possibly devastating global warming trend that could flood coastal cities, disrupt the food chain, and change climate patterns significantly, many states have taken the initiative and enacted renewable energy portfolios to fund the transition to renewable energy resources. Many remaining states are in the process of developing their own renewable energy portfolios. These renewable energy portfolios provide significant state- and utility-sponsored financial incentives for the commercial, industrial, and residential use of renewable energy systems and fuels. On the city level, many mayors from major cities around the world have made commitments to cut greenhouse gas emissions to slow the rate of global warming. Many of these cities are coastal and could be severely impaired or destroyed from rising sea levels from global warming. So, civic action to switch to cleaner energy options is beginning in earnest.

The Future Energy:

Hydrogen (where the hydrogen is derived from renewable energy sources), ethanol, biodiesel, and other forms of renewable fuels are on their way in. Direct and indirect conversion of solar energy, including wind, biomass, wave/tidal power, and small-scale hydroelectric power will increasingly be part of the energy infrastructure that energy engineers will design and build. Nuclear power emissions are clean, but the nuclear power industry has significant obstacles such as storage of radioactive wastes for many thousands of years. In addition, there are security concerns to safeguard radioactive material from being stolen for production of atomic weapons. The bottom line with energy is that it needs to be relatively non polluting and indefinitely available. It is a very dynamic time for energy engineers as the entire, world-wide energy picture transitions to clean renewable technologies. This will eventually add a lot of stability to the world economy, the world political environment and to everyone's lives. The stability will come from the fact that renewable energy technologies can be used to tap the natural energy resources that are available everywhere. Stability will also come, as the environment regenerates, the climate stabilizes and resources remain available for our sustenance.

- Energy resources are available to supply the world's expanding needs without environmental detriment.
- Ethical principles seem increasingly likely to influence energy policy in many countries, which augurs well for nuclear energy.

- The competitive position of nuclear energy "is robust from a sustainable development perspective since most health and environmental costs are already internalized.

Until about 20 years ago, energy sustainability was thought of simply in terms of availability relative to the rate of use. Today, in the context of the ethical framework of sustainable development, including particularly concerns about global warming, other aspects are equally important. These include environmental effects and the question of wastes, even if they have no environmental effect. Safety is also an issue, as well as the broad and indefinite aspect of maximizing the options available to future generations.

Sources of Energy:

Harnessing renewable energy such as wind and solar is an appropriate first consideration in sustainable development, because apart from constructing the plant, there is no depletion of mineral resources and no direct air or water pollution. In contrast to the situation even a few decades ago, we now have the technology to access wind on a significant scale for electricity. But harnessing these 'free' sources cannot be the only option. Renewable sources other than hydro – notably wind and solar – are diffuse, intermittent, and unreliable by nature of their occurrence. These sources are intrinsically unsuited to meeting the demand for continuous, reliable supply on a large scale – which comprises most demand in developed countries.

These aspects offer a technological challenge of some magnitude, given that electricity cannot be stored on any large scale. For instance, solar-sourced electricity requires collecting energy at a peak density of about 1 kilowatt (kW) per square meter when the sun is shining to satisfy a quite different kind of electricity demand – one which mostly requires a relatively continuous supply.

Wind is the fastest-growing source of electricity in many countries, albeit from a low base, and there is a lot of scope for further expansion. While the rapid expansion of wind turbines in many countries has been welcome, capacity is seldom more than 30% utilized over the course of a week or year, a consequence of the unreliability of the source and the fact that it does not and cannot match the pattern of demand. The rapid expansion of wind farms is helped considerably by generous government-mandated grants, subsidies and other arrangements ultimately paid by consumers. But there is often a strong groundswell of opposition on aesthetic grounds from the countryside where the turbines are located.

A fuller treatment of electricity from renewable sources is in the information page on Renewable Energy and Electricity.

Apart from renewable, it is a question of what is most abundant and least polluting. Today, to a degree almost unimaginable even 25 years ago, there is known to be an abundance of many energy resources in the ground. Coal and uranium (not to mention thorium) are available and unlikely to be depleted this century.

The criteria for any acceptable energy supply will continue to be cost and safety, as well as environmental considerations. Addressing environmental effects usually has cost implications, as the current greenhouse debate makes clear. Supplying low-cost electricity with acceptable safety and low environmental impact will depend substantially on developing and deploying reasonably sophisticated technology. This includes both large-scale and small-scale nuclear energy plants, which can be harnessed directly to industrial processes such as hydrogen production or desalination, as well as their traditional role in generating electricity.

Generally 'renewable' relates to harnessing energy from natural forces which are renewed by natural processes, especially wind, waves, sun and rain, but also heat from the Earth's crust and mantle.

Although it shares many attributes with technologies harnessing these natural forces – for instance radioactive decay produces much of the heat harnessed geothermally – nuclear power is usually categorized separately from 'renewables'.

Conventional nuclear power reactors do use a mineral fuel and demonstrably deplete the available resources of that fuel. In such a reactor, the input fuel is uranium-235 (U-235), which is part of a much larger mass of uranium – mostly U-238. This U-235 is progressively 'burned' to yield heat. But about one-third of the energy yield comes from something which is not initially loaded in: plutonium-239 (Pu-239), which behaves almost identically to U-235. Some of the U-238 turns into Pu-239 through the capture of neutron particles, which are released when the U-235 is 'burned'. So the U-235 used actually renews itself to some extent by producing Pu-239 from the otherwise waste material U-238.

This process can be optimized in fast neutron reactors, which are likely to be extensively deployed in the next generation of nuclear power reactors. A fast neutron reactor can be configured to 'breed' more Pu-239 than it consumes (by way of U-235 + Pu-239), so that the system can run indefinitely. While it can produce more fuel than it uses, there does need to be a steady input of reprocessing activity to separate the fissile plutonium from the uranium and other materials discharged from the reactors. This is fairly capital-intensive but well-proven and straightforward. The used fuel from the whole process is recycled and the usable part of it increases incrementally.

As well as utilizing about 60 times the amount of energy from uranium, fast neutron reactors will unlock the potential of using even more abundant thorium as a fuel (see information page on Thorium). In addition, some 1.5 million tons of depleted uranium now seen by some people as little more than a waste, becomes a fuel resource. The consequence of this is that the available resource of fuel for fast neutron reactors is so plentiful that under no practical terms would the fuel source be significantly depleted.

Regardless of the various definitions of 'renewable', nuclear power therefore meets every reasonable criterion for sustainability, which is the prime concern.

Wastes: Burning fossil fuels produces primarily carbon dioxide as waste, which is inevitably dumped into the atmosphere. With black coal, approximately one tons of carbon dioxide results from every thousand kilowatt hours generated. Natural gas contributes about half as much CO_2 as coal from actual combustion, and also some (including methane leakage) from its extraction and distribution. Oil and gas burned in transporting fossil fuels adds to the global total. As yet, there is no satisfactory way to avoid or dispose of the greenhouse gases which result from fossil fuel combustion.

Nuclear wastes: Nuclear energy produces both operational and decommissioning wastes, which are contained and managed. Although experience with both storage and transport over half a century clearly shows that there is no technical problem in managing any civil nuclear wastes without environmental impact, the question has become political, focusing on final disposal. In fact, nuclear power is the only energy-producing industry which takes full responsibility for all its wastes, and costs this into the product – a key factor in sustainability.

Ethical, environmental and health issues related to nuclear wastes are topical, and their prominence has tended to obscure the fact that such wastes are a declining hazard, while other industrial wastes retain their toxicity indefinitely.

Regardless of whether particular wastes remain a problem for centuries or millennia or forever, there is a clear need to address the question of their safe disposal. If they cannot readily be destroyed or denatured, this generally means that they need to be removed and isolated from the biosphere. This may be permanent, or retrievable.

An alternative view asserts that indefinite surface storage of high-level wastes under supervision is preferable. This may be because such materials have some potential for recycling as a fuel source, or because progress towards successful geological disposal would simply encourage continued use and expansion of nuclear energy. However, there is wide consensus that dealing effectively with wastes to achieve high levels of safety and security is desirable in a 50-year perspective, ensuring that each generation deals with its own wastes.

Sustainable development Criteria:

"Sustainable development is development that meets the needs of the present without compromising the ability of future generations to meet their own needs."

- ✓ Balance of flows
- ✓ Cycling of matter
- ✓ Ecosystems integrity
- ✓ Social system integrity

Balance of flows *(system condition 1)*

In order for a society to be sustainable, nature's functions and diversity are not systematically subject to increasing concentrations of substances extracted from the Earth's crust.

- Living off current solar income, not the principal
- This criteria covers global climate change and other potential problems such as acid rain, due to the accumulation of fossil fuels and their by products in the biosphere.
- Metals and other minerals must not be extracted at a faster rate than their re-deposit and reintegration into the earth's crust. Heavy metals, such as lead, mercury and cadmium, must not systematically accumulate in the biosphere.

Cycling of matter *(system condition 2)*

In order for a society to be sustainable, nature's functions and diversity are not systematically subject to increasing concentrations of substances extracted from the Earth's crust.

- No reliance on persistent synthetic compounds, or human-made substances that cannot be broken down and cycled in natural systems
- Human made substances not be produced at a rate faster than their assimilation in the biosphere.

Ecosystem integrity *(system condition 3)*

In order for a society to be sustainable, nature's functions and diversity are not systematically impoverished by overharvesting or other forms of ecosystem manipulation.

The productive surfaces of the biosphere must not be diminished in quality or quantity.

- We must not harvest more from the biosphere than can be recreated and renewed.
- Biodiversity and diversity of habitats are essential for ecosystem integrity.

Social system integrity *(system condition 4)*

In order for a society to be sustainable, resources are used efficiently and fairly in order to meet basic human needs worldwide.

- integrate and enhance economic vitality and social equity

Review Questions

1. Discuss the energy systems in past and present? Cogitate your ideas for generation of energy in future?

2. Explain in detail the energy systems in past and present. What ideas do you have for generation of power in future by overcoming the present ill effects Explain?

3. What is sustainable development? On what energy technologies we need to depend for the sustainable development and discuss.

4. Explain the criteria for the sustainable development of the energy systems?

Objective Type Questions

1. Best example for sustainable energy production

    a) Nuclear   b) Gas power   c) coal   d) Diesel power

2. Which is the most important factor in Nuclear power generation?

    a) Green house gas emission   b) Waste disposal
    c) High fuel cost   d)none of the above

CHAPTER 9

ENERGY MANAGEMENT AND PLANNING

This Chapter has received attention over a period of many years. First there is a need for a strong corporate-wide energy management program that is fully supported by management and all other employees. Second, it is required that equipment such as boilers, compressors, and pumps, which are common to most processing plants. should be carefully reviewed to ensure that they are being operated as efficiently as possible, and also reviewed in the light of the introducing new more efficient replacements. Thirdly techniques are required to be adopted that allow fine-tuning or optimization of operations to produce additional savings.

9.1 Management of Energy

India is a Developing nation. Its per capita Energy Consumption is very low. To achieve Economic Growth, we need to & have to use more & more energy to increase the pace of development. We need to increase the manufacturing of good in Quality & Volume. It is estimated that Industrial energy use in developing countries constitutes about 45-50 % of the total commercial energy consumption. Much of this energy is converted from imported oil, the price of which has increased tremendously so much so that most of developing countries spent more than 50 % of their foreign exchange earnings. Not with standing these fiscal constraints, developing countries need to expand its industrial base like us if it has to generate the resources to improve the quality of life of its people. The expansion of industrial base does require additional energy inputs which become more &more difficult in the present scenario.

Generation of power needs resources. Resources available on earth are of Diminishing Nature. It is getting depleted very fast with time as use is increasing exponentially. There are some resources, which are Renewable e.g. Solar Power, Wind Power and Geothermal Power. Technology is also being developed to harness these Renewable Resources to generate Power. The capital investment requirement is very high as compared to normally available resources. It can be quoted here that with the available technology, we could hardly generate 5% of total power generation as on date .Hence, to restrict the use or increase the life of diminishing type of resources.

The introduction of a company-wide supported energy management program has been frequently demonstrated as the ideal and most cost-effective way to bring about energy efficiency improvements. The introduction of such a program produces a culture of improvement that is supported by all employees and allows the introduction of carefully formulated guidelines which allow for the efficient management of energy use within the company or group of companies. Without this program and set of guidelines then energy management is likely to be haphazard and produces a sense of confusion in the workforce. Areas of confusion are quite often linked to a lack of properly organized communication systems which are needed to implement a program of energy efficiency changes. Considering the opportunities that usually exist in the improvement of energy efficiency in many energy-using processes, it is surprising that there are still barriers that prevent the adoption of wide-scale improvements in energy management within organizations. Worrell and Galitsky with the U.S. EPA. through ENERGY STAR have worked with many of the leading industrial manufacturers to identify the basic aspects of an effective energy management program.

In order to set up an effective energy management program, there is a need to establish an energy director who can oversee and manage the supported program. This can be followed by the appointment of an energy team, which can then establish the required procedures needed for such actions to assess performance regularly review the energy information that is available and assess the need of additional data collection, assess any technical requirements that are needed, and carry out some form of benchmarking. Using this type of management system in an organization is able to develop an initial baseline of the current performance related to energy use and set goals for improvement.

Let us see the other aspect of life, whereas everybody can't understand all technical reasons or benefits of the whole world until he himself realizes some benefit for his action or efforts. In this competitive world, cost competitiveness is very essential for survival of every individual. To establish any work / motive or task, energy in one or other form is an essential component. Thus the need to conserve energy, particularly in industry and commerce is strongly felt as the energy cost takes up substantial share in the overall cost structure of the operation.

Hence it calls Management of Energy or in other words Management of Resources or Energy Conservation. It becomes clear from the above data & statement that Energy needs to be Managed / Resources needs to be managed irrespective of a developed nation or developing nation.

If current performance is correctly established, then the setting goals is obviously much easier, and this in turn allows the implementation plan that can provide the necessary support to achieve these goals. The success of this performance improvement cannot be made unless all employees of the organization are informed and involved. There is usually a need for staff training that can be carried out in-house, or outside experts can also be brought in for this. The level of this support helps to cement the idea that improvement in energy efficiency is a concern for all members of staff: Communication to staff on how the improved performance is progressing is vitally important. The progression of performance toward the initial goals needs to be accessed and communicated, and any best practices that have been identified should be implemented as widely as possible.

The major elements are depicted in below Figure.

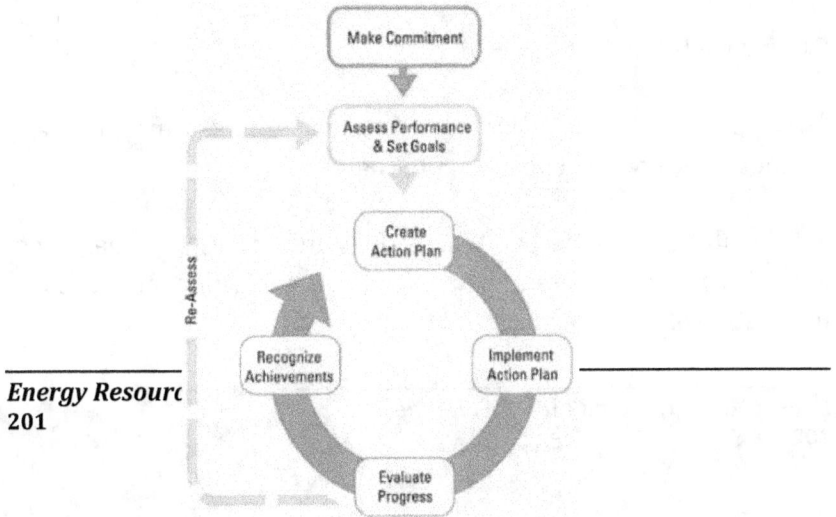

Figure 9.1 Energy Management Overview
(Source: Guidelines for Energy Management,U.S. EPA EnergyStar Program)

STEP 1: Make Commitment

Organizations seeing the financial returns from superior energy management continuously strive to improve their energy performance. Their success is based on regularly assessing energy performance and implementing steps to increase energy efficiency.

No matter the size or type of organization, the common element of successful energy management is commitment. Organizations make a commitment to allocate staff and funding to achieve continuous improvement.

To establish their energy program, leading organizations form a dedicated energy team and institute an energy policy.

Form a Dedicated Team

1.1 Appoint an Energy Director - Sets goals, tracks progress, and promotes the energy management program.

1.2 Establish an Energy Team - Executes energy management activities across different parts of the organization and ensures integration of best practices.

Institute an Energy Policy

1.3 Institute an Energy Policy - Provides the foundation for setting performance goals and integrating energy management into an organization's culture and operations.

STEP 2: Assess Performance

Understanding current and past energy use is how many organizations identify opportunities to improve energy performance and gain financial benefits. Assessing performance is the periodic process of evaluating energy use for all major facilities and functions in the organization and establishing a baseline for measuring future results of efficiency efforts.

Key aspects include:

Data Collection and Management
2.1 Gather and track data - Collect energy use information and document data over time.
Baselining and Benchmarking
2.2 Establish baselines - Determine the starting point from which to measure progress.

2.3 Benchmark - Compare the energy performance of your facilities to each other, peers and competitors, and over time to prioritize which facilities to focus on for improvements.

Analysis and Evaluation

2.4 Analyze - Understand your energy use patterns and trends.

2.5 Technical assessments and audits - Evaluate the operating performance of facility systems and equipment to determine improvement potential.

Assessing your energy performance helps:

- Categorize current energy use by fuel type, operating division, facility, product line, etc.
- Identify high performing facilities for recognition and replicable practices.
- Prioritize poor performing facilities for immediate improvement.
- Understand the contribution of energy expenditures to operating costs.
- Develop a historical perspective and context for future actions and decisions.
- Establish reference points for measuring and rewarding good performance.

STEP 3: Set Goals

Performance goals drive energy management activities and promote continuous improvement. Setting clear and measurable goals is critical for understanding intended results, developing effective strategies, and reaping financial gains. Well-stated goals guide daily decision-making and are the basis for tracking and measuring progress. Communicating and posting goals can motivate staff to support energy management efforts throughout the organization.

The Energy Director in conjunction with the Energy Team typically develops goals.

To develop effective performance goals:

3.1 Determine scope - Identify organizational and time parameters for goals.

3.2 Estimate potential for improvement - Review baselines, benchmark to determine the potential and order of upgrades, and conduct technical assessments and audits.

3.3 Establish goals - Create and express clear, measurable goals, with target dates, for the entire organization, facilities, and other units.

Setting goals helps the Energy Director

- Set the tone for improvement throughout the organization
- Measure the success of the energy management program
- Help the Energy Team to identify progress and setbacks at a facility level
- Foster ownership of energy management, create a sense of purpose, and motivate staff
- Demonstrate commitment to reducing environmental impacts
- Create schedules for upgrade activities and identify milestones

STEP 4: Create Action Plan

With goals in place, your organization is now poised to develop a roadmap to improve energy performance. Successful organizations use a detailed action plan to ensure a systematic process to implement energy performance measures. Unlike the energy policy, the action plan is regularly updated, most often on an annual basis, to reflect recent achievements, changes in performance, and shifting priorities. While the scope and scale of the action plan is often dependent on the organization, the steps below outline a basic starting point for creating a plan.

4.1 Define technical steps and targets

4.2 Determine roles and resources

Get buy-in from management and all organizational areas affected by the action plan before finalizing it. Work with the Energy Team to communicate the action plan to all areas of the organization.

STEP 5: Implement Action Plan

People can make or break an energy program. Gaining the support and cooperation of key people at different levels within the organization is an important factor for successful action plan implementation in many organizations. In addition, reaching your goals frequently depends on the awareness, commitment, and capability of the people who will implement the projects.

To implement your action plan, consider taking the following steps:

5.1 Create a communication plan - Develop targeted information for key audiences about your energy management program.

5.2 Raise awareness - Build support all levels of your organization for energy management initiatives and goals.

5.3 Build capacity - Through training, access to information, and transfer of successful practices, procedures, and technologies, you can expand the capacity of your staff.

5.4 Motivate - Create incentives that encourage staff to improve energy performance to achieve goals.

5.5 Track and monitor - Using the tracking system developed as part of the action plan to track and monitor progress regularly.

STEP 6: Evaluate Progress

Evaluating progress includes formal review of both energy use data and the activities carried out as part of the action plan as compared to your performance goals.

Evaluation results and information gathered during the formal review process is used by many organizations to create new action plans, identify best practices, and set new performance goals.

Key steps involved include:

6.1 Measure results - Compare current performance to established goals.

6.2 Review action plan - Understand what worked well and what didn't in order to identify best practices.

Regular evaluation of energy performance and the effectiveness of energy management initiatives also allows energy managers to:

- Measure the effectiveness of projects and programs implemented
- Make informed decisions about future energy projects
- Reward individuals and teams for accomplishments
- Document additional savings opportunities as well as non-quantifiable benefits that can be leveraged for future initiatives.

STEP 7: Recognize Achievements

Providing and seeking recognition for energy management achievements is a proven step for sustaining momentum and support for your program.

Providing recognition to those who helped the organization achieve these results motivates staff and employees and brings positive exposure to the energy management program.

Receiving recognition from outside sources validates the importance of the energy management program to both internal and external stakeholders, and provides positive exposure for the organization as a whole.

Key steps in providing and gaining recognition include:

7.1 Providing internal recognition - to individuals, teams, and facilities within your organization.

7.2 Receiving external recognition - from government agencies, the media, and other third party organizations that reward achievement.

How to Manage Energy Systems?

Energy management is not by chance / incident / accident. It is a Mission with a Target. It can't be done single handedly or by sitting on a table. It needs coordinated effort by team of energy conscious people with a milestone to be established. Very concerted efforts in a planned manner to established Energy Management. Strategy needs to be established based on the Target of Energy Conservation.

Strategy / Methodology of Energy Management:

Having established the need of Energy Management /Conservation. A systematic approach needs to be discussed and concluded. Same of steps to reach to the target of Energy Conservation can be listed as below: -

1) Identification of Inefficient areas / Equipments: -
   Enlistment or knowledge of type of energy being used.
   Study of machines / Technology employed.
   Process study and identification of major energy consumption areas.

In depth process study to identify the inefficient use of energy.

2) Identification of Technology / Equipment requirement.

3) Discussion, Brain storming & Conclusion of resources requirement.

4) Management of resources like Manpower, Machine or Technology.

5) Evaluate your actions / efforts to estimate the Rate of Return "Inefficient action / efforts cannot give efficient results" "Only Efficient efforts & Economic ideas need to Betested."

6) Implementation of New Process / New Technology / New Machines.

7) Re-evaluate your actions / Your Efforts.

Energy Management Techniques:

1) Self knowledge & Awareness among the masses
2) Re-engineering & evaluation
3) Technology Upgradations

1. Self Knowledge & Awareness among the Masses:

For the successful Energy management & implementation, the knowledge of process & machine for the leader is very important. On the first instance, there is always a resistance from the user. There might be psychological mind blocks in the user's mind. This needs to be made known & clarified. It is further more important to make the owner of the process understand the cost benefit of the energy conservation. Creating Awareness to the process owner can give most economic & low cost solutions to save energy. We have realized about 5 % of energy saving just because of Awareness of the people. e.g:

A. The concept of "Zero Production = Zero Power consumption".

People realized this concept and they started switching off the Auxiliaries during Idling of main machines.

B. Later on, we introduced microprocessor-based timer to auto switch off auxiliary equipments during Idle period of machines.

2. Re-engineering and Technology Up gradations:

After utilizing the low cost or awareness concept, we need to do in depth study the process /machine. We need to ascertain, the scope & extent of Energy Conservation in the area under consideration. Evaluate the existing situation / employed technology in terms of process requirements &production capacity & capability. Sometimes, we do land into a situation of handicap with machine capacity & capability for the sake of Energy Conservation. It must not be done. Once it is established, that there is a potential of energy optimization. We need to start evaluation& re-engineering of the process / equipment. It may be terms of layout, motor capacity, types of starters employed, nature of loads etc.

Technology Upgradations: After having established the scope of energy conservation in the specified area. The latest technology availability is suitability, sustainability & pricing needs to be studied. Economics needs to be worked out like Pay back period, Return of Investment, Quality of energy savings etc.

Please remember "BETTER THE DIAGNOSIS, BEST WILL BE THE RESULT"

9.2 Energy Systems Planning & Control

Energy is essential for all human activities, and its availability is critical to economic and social development. Energy is the engine for the production of goods and services across all economic sectors. It is vital to the provision of basic civic services in education, health care, clean water supply and sanitation, and also for wealth creation. Lack of energy is a contributing factor to the poverty of individuals, communities, nations and regions. While not an end itself, energy, jointly with appropriate technologies and infrastructure, generates the services modern societies demand (transportation, lighting, air conditioning, information exchange, etc.).

Energy planning has a number of different meanings. However, one common meaning of the term is the process of developing long-range policies to help guide the future of a local, national, regional or even the global energy system. Energy planning is often conducted within Governmental organizations but may also be carried out by large energy companies such as electric utilities or oil and gas producers.

Energy planning may be carried out with input from different stakeholders drawn from government agencies, local utilities, academia and other interest groups. Energy planning is often conducted using integrated approaches that consider both the provision of energy supplies and the role of energy efficiency in reducing demands. Energy planning should always reflect the outcomes of population growth.

Energy planning has traditionally played a strong role in setting the framework for regulations in the energy sector (for example, influencing what type of power plants might be built or what prices were charged for fuels). But in the past two decades many countries have deregulated their energy systems so that the role of energy planning has been reduced, and decisions have increasingly been left to the market. This has arguably led to increased competition in the energy sector, although there is little evidence that this has translated into lower energy prices for consumers. Indeed in some cases, deregulation has led to significant concentrations of "market power" with large very profitable companies having a large influence as price setters.

This trend now seems to be reversing as concerns grow over the environmental impacts of energy consumption and production, particularly in light of the threat of global climate change, which is caused largely by emissions of greenhouse gases from the world's energy systems. Many OECD countries and some U.S. states are now moving to more closely regulate their energy systems. For example, many countries and states have been adopting targets for emissions of CO_2 and other greenhouse gases. In light of these developments, it seems likely that integrated energy planning will become increasingly important.

A new trend in energy planning known as Sustainable Energy Planning takes a more holistic approach to the problem of planning for future energy needs. It is based on a structured decision making process based on six key steps, namely:

1. Exploration of the context of the current and future situation

2. Formulation of particular problems and opportunities which need to be addressed as part of the Sustainable Energy planning process. This could include such issues as "Peak Oil" or "Economic Recession/Depression".

3. Create a range of models to predict the likely impact of different scenarios. This traditionally would consist of mathematical modeling but is evolving to include "Soft System Methodologies" such as focus groups, peer ethnographic research, "what if" logical scenarios etc.

4. Based on the output from a wide range of modeling exercises and literature reviews, open forum discussion etc., the results are analyzed and structured in an easily interpreted format.

5. The results are then interpreted in order to determine the scope, scale and likely implementation methodologies which would be required to ensure successful implementation.

6. This stage is a quality assurance process which actively interrogates each stage of the Sustainable Energy Planning process and checks if it has been carried out rigorously, without any bias and that it furthers the aims of sustainable development and does not act against them.

7. The last stage of the process is to take action. This may consist of the development, publication and implementation of a range of policies, regulations, procedures or tasks which together will help to achieve the goals of the Sustainable Energy Plan.

Designing for implementation is often carried out using "Logical Framework Analysis" which interrogates a proposed project and checks that it is completely logical, that it has no fatal errors and that appropriate contingency arrangements have been put in place to ensure that the complete project will not fail if a particular strand of the project fails.

Sustainable energy planning is particularly appropriate for communities who want to develop their own energy security, while employing best available practice in their planning processes.

Energy Systems Control:

Energy use can he controlled in order to reduce costs and maximize profits. The controls can be as simple as manually turning off a switch, but often automated controls ranging from simple clocks to sophisticated computers are required. Our view is that the control should be as simple and reliable as possible. Consequently, this chapter starts with manual controls and proceeds through timers, programmable controllers, and digital computers. As one moves through this hierarchy of controls, each level of automation and complexity requires additional expenditure of capital. That is, the automated controls are more expensive, but they do more. Because choosing the proper type of control is often a difficult task, we will explore this decision process.

Computers can also help the energy manager in the analysis of proposed and present energy systems. Some excellent large-scale computer simulation programs have been written that enable the energy analyst to try alternative scenarios of energy equipment and controls, so in the last part of this chapter we discuss these computer programs and their use. BLAST 3.0 and DOE-2.1D are the two described in depth, but several others are mentioned.

Every piece of energy-consuming equipment has some form of control system associated with it. Lights have on-off wall switches or panel switches, and some have timers and dimmer controls. Motors have on-off switches, and some have variable speed controls. Air conditioners have thermostats and fan switches; they sometimes have night setback controls or timers. Large air conditioning systems have extensive controls consisting of several thermostats, valve and pump controls, motor speed controls, and possibly scheduling controls to optimize the operation of all of the components. Water and space heaters have thermostats and pump controls or fan motor controls. Large heating systems have modulating controls on the boilers and adjustable speed drives on pumps and variable air volume fans.

These controls are necessary for the basic safety of the equipment and the operators, as well as for the proper operation of the equipment and systems. Our interest is in the energy consumption and energy efficiency of this equipment and these systems, and the controls have a significant impact on both these areas. Controls allow unneeded equipment to be turned off, and allow equipment and systems to be operated in a manner that reduces energy costs. This may include reductions in the electric power and energy requirements of equipment, as well as the power and energy requirements associated with other forms of energy such as oil, gas and purchased steam.

Types of controls:

In this section, we present the different types of controls in order of increasing complexity and cost. In each subsection, the control discussed can perform the functions covered in that subsection as well as all those functions covered in the preceding subsections. For example, the functions discussed in the second subsection on timers can be performed very well by a timer or any of the succeeding types of controls (programmable controllers, microprocessors, and large computers) hut not by a manual system.

Manual Systems:

Manual control systems can be used to turn equipment off when it is not needed. Turning equipment off when not in use can lead to dramatic savings. For example, lights are often left on at night, but they should normally be turned off whenever possible. (Often a small series of lights is left on for security purposes.)

One of the best opportunities for manual control exists in the area of exhaust and makeup air fans. These fans are often located at the top of a high ceiling, and they are frequently left on unnecessarily because their running is undetectable without close scrutiny. The savings for turning off exhaust fans is twofold. First, electricity is no longer required to run the fan motor, and, second, conditioned air is no longer being exhausted.

Basic Automatic Controls—Timers and Dimmers:

The next step in level of control complexity is the use of automatic controls such as timers and dimmers. Timers can range from very simple clocks to fairly complicated central time clocks with multiple channels for controlling numerous pieces of equipment on different time schedules. Automatic timer controls can range from simple thermostats each with a built-in time clock (costing somewhere around $100 each) to a central time clock that overrides all the thermostats. An installed single-channel central time clock will cost around $1000, hut it can control numerous thermostats if all are on the same schedule. Different setback schedules require multiple channels, increasing the cost somewhat.

Some companies have utilized time clocks to duty cycle equipment such as exhaust fans. For example, a large open manufacturing area will likely have several exhaust fans. If there are six fans, then a central time clock could turn one fan off each 10 minutes and rotate so that each fan is off 10 minutes of each hour, but no more than one fan is ever off at the same time. This saves on electrical consumption (kWh) to run the fan, electrical demand (kW, since one fan is *off* at any time), and heating (since less conditioned air is exhausted). General ventilation over a wide area is maintained. Of course, care must be taken to ensure that no ventilation problems develop.

The use of timers allows a company to start-stop equipment at exactly the correct time. It is not necessary to wait for maintenance people to make their rounds, turning off equipment and adjusting thermostats. However, although timers don't forget to do their job, they do suffer from other problems. For example, power outages may require timers to be reset unless a battery backup is used. Also, arrival or termination of daylight savings time requires all timers to be set up an hour in the spring and back an hour in the fall. Finally, the clocks must be maintained and replaced as they wear out.

The authors had an opportunity to audit a plant that had sophisticated time clock controls on its equipment, but management was not maintaining the clocks. The 7-day time clocks allowed for night and weekend setbacks. The audit was done on a Thursday, but the time clocks read Saturday. Consequently, the thermostats were on night setback, and the employees were cold. To remedy this, maintenance had purchased several additional portable heaters. If they had come in on a Saturday, when the clock read Monday, the plant would have been nice and warm. In this case, the poorly maintained clocks were costing the company a great deal of money. Timers and any other type of control system must be maintained.

Another type of control that has some attractive savings potential is a light dimmer. Dimmers can be automatically controlled depending on time, and on natural lighting levels if photocell sensors are used. it is important to be sure the dimming system chosen actually reduces electrical consumption and is not simply a rheostat (variable resistor) that consumes the same amount of energy regardless of the amount of light delivered. Supermarkets can often use relatively sophisticated dimming systems.

For example, supermarkets might:

1. Use photocells to detect natural light and dim the window lights as appropriate.
2. Use photocells that turn parking and security lights off at dawn arid on at dusk.
3. Use photocells to determine dusk so that interior lighting can be reduced. (Studies have shown that people coming from a dark street to a brightly lit room are actually uncomfortable. Lower lighting levels are preferred.)

9.3 Integrated Energy Planning

Put very simply, integrated energy planning involves estimating how much energy all the different consumers (e.g. industry or households) will need in the future to deliver certain services and then identifying a mix of appropriate sources and forms of energy to meet these energy service needs in the most efficient and socially beneficial manner.

Integrated energy planning is much more than just energy planning. Energy planning has traditionally been done by energy companies to work out their company strategy. Since this was aimed at increasing their share of the market and making profit, energy planning only considered the economic benefits to the company, without taking into account the nature of users real needs, nor the macro-economic', social and environmental consequences of different options.

In contrast, integrated energy planning allows for a departure from business as usual, and has three key requirements:

- ✓ The inclusion of all energy service needs and supply side options, including energy savings and efficiency interventions

- ✓ The inclusion of all costs and benefits (economic, social and environmental). including long-term benefits and costs and macro-economic impacts (e.g. economic competitiveness in 20 years) in describing possible futures (scenarios) for the entire energy sector

- ✓ Setting goals for the future, based on a description of the most desirable situation at the end of the planning horizon, that embrace the whole energy system and its impacts

Necessity of integrated energy planning:

The objective of integrated energy planning is to decide how to meet energy service needs in the most efficient and socially beneficial manner, keeping control of economic costs while also serving national imperatives such as job creation and poverty alleviation. It also allows for consideration of substitution between energy carriers, e.g. reducing electricity demand through mandatory introduction of solar water heating or fuel switching away from imported oil.

Phases of integrated energy planning:

There are three main phases to integrated energy planning:

Phase 1: Reference Energy System and how it evolved

The foundation for energy planning is a comprehensive description of the whole energy system for the country (or city', or region), as it exists at present. This is known as the reference energy system (RES). Due to the challenges of data collection and processing. the RES will generally be two or three years in the past by the time it is completed.

- primary energy supply (e.g. oil, coal, hydro; including imports and exports)
- transformation to secondary energy (electricity generation, oil refining, coal to gas and liquid fuel etc.)
- transport and distribution (including a reflection of system losses) &
- Final consumption, per sector (e.g. iron & steel, mining and quarrying, rail transport, residential).

This information is presented for all the energy carriers, both primary sources: coal, oil, natural gas, nuclear, hydro, solar, geothermal, wind; and secondary carriers: petroleum products. A complete energy balance will reflect losses in transformation, such as approximately 65% of energy lost in thermal electricity power plants (coal and nuclear), energy used in mining coal and consumed by utilities, as well as losses in distribution. The RES should also take account of the modes of final consumption, in other words differentiate between final energy as delivered to customers, (electricity, petrol, gas) and useful energy, i.e. the energy output (light, heat, movement) of the end-use appliance (e.g. lamp, stove, geyser, pump or car). This is important because the ultimate objective of the energy system is to deliver energy services (e.g. heat for cooking, water heating and industrial processes or mechanical power for processing. manufacture and transport).

In addition to this 'snapshots of the energy system, we need information on the trends and dynamics that led to the current situation (the video as well as the latest snapshot). This will include data showing the drivers of supply and demand (including economic growth, international markets and political agendas) and how they interacted for example if rising oil prices led to increased energy efficiency. as in the USA in the early eighties, or high prices and sanctions leading to an emphasis on energy security in South Africa, sufficient to justify massive subsidies for coal to liquid fuel and electricity plants.

Phase 2: Energy forecasting and scenarios

Energy forecasting involves using the snapshot and video taken in phase Ito map possible evolutions of the energy system. The first step is to choose a time horizon for the planning exercise. The energy forecast must then describe the evolution of the reference energy system from the base year to the horizon year. In doing this the energy forecast consists of two parts:

- Future energy demand
- Future energy supply to meet that demand (from sources through to useful energy)

Energy forecasting does not simply mean predicting the future based on business as usual. It should also take account of 'suppressed demands (needs that are not expressed through purchasing power e.g. services for the poor) and the potential for changes in market conditions e.g. commodity prices and pollution charges. Thus forecasting will indicate a range for future energy demand under different assumptions.

Scenarios then examine the conditions under which future energy demand can be met in ways that are most beneficial socially, economically and environmentally. By considering which systems would be best for the future, and not simply assuming business as usual, we can then decide what policies (for example increasing the current renewable energy target of less than 1 % by 2013 to 20% by 2020) and strategies (once off subsidies versus feed-in-tariffs to support the introduction of renewable energy into the power market) are needed to transform the energy system to best serve the needs of society as a whole, rather than simply increasing sales or profits for business on the supply side. To help determine what is viable (within given constraints such as resource availability) and how policy instruments are likely to impact on the energy system and beyond many models are available. These models produce different scenarios or possible futures. While these models are very useful in energy forecasting. they are not perfect. Rather they're tools to test the impacts of possible strategies and policies.

Phase 3: Planning

With these different future energy scenarios in mind, the actual plans to reach the best possible future can be drawn up. it's important that policy makers are involved in this phase, as they are the people responsible for ensuring that the good of South Africa's society as a whole is prioritized over sectoral interests. So for example, policy makers may well decide to pass laws to support the introduction of renewable energy because thousands of jobs will be created, even if in the short term this choice is more expensive than business as usual. Similarly parliament may call for mandatory standards for energy efficiency and conservation.

9.4 Role of Institutions in Managing Energy systems

Among the agencies that oversee energy policy in India are the Ministry of Petroleum and Natural Gas, the Ministry of Coal, The Ministry of Non Conventional Energy Sources, The Ministry of Environment and Forests, the Department of Atomic Energy, and the Ministry of Power. Within the Ministry of Power, the Central Electricity Regulatory Commission (CERC) works closely with individual SEBs and utilities in power generation, T&D of electricity. Under the Department of Atomic Energy the Atomic Energy Commission was established as the policy making body for the development and utilization of atomic energy for peaceful purposes. The ministries that oversee transportation of fuels arc also important. The Ministry of Shipping Transport is responsible for the importation of energy aboard ships of the state owned Shipping Corporation of India. The Parliamentary Committee on Energy and the Energy Policy Division of the Planning Commission also are involved in steering *policies* concerning energy.

India's energy sector is almost totally in the control of the government. The publicly owned Coal India Ltd. and its seven subsidiaries produce more than 90 per cent of India's coal. Through time Ministry of Coal, the Planning Commission and the Public Investment Board, the government controls all facets of coal operations. In addition, the Ministry of Coal oversees the Neyveli Lignite Corporation (NLC). Singareni Collieries and the Coal Controllers Organization which was established to monitor coal quality and distribution.

In the petroleum sector, with the exception of two joint refining ventures the government owns and manages the entire oil and gas activities. The Oil and Natural Gas Commission (ONGC) accounts for roughly 90 per cent of total crude oil upstream production and Oil

India (OIL) produces the remaining 10 per cent. While six separate companies handle the refining and marketing of petroleum products, the Indian Oil Corporation (IOC) carries out all refining and distribution operations and is the main importer of crude oil and refined products. Most gas distribution is owned and managed by the Gas Authority of India Ltd (GAIL).

In the power sector, both the central and state governments share responsibility for supplying electricity. Since the 1970s, the government of India (GOI) has been actively involved in power development and establishing its utilities which complement state efforts. GOI currently owns (among others) the National Thermal Power Corporation (NTPC), National Hydroelectric Power Corporation NHPC), the Nuclear Power Corporation (NPC) and the Power Grid Corporation of India (POWERGRID) a transmission company and grid operator. As amended, the National Electricity Supply Act of 1948 puts the Central Electricity Authority (CEA) in charge of national power policy and planning, coordinating and regulating sector development. Through the Ministry of Power (MOP) the government approves national power plans and makes rules for carrying out CEA functions.

India's largest public petroleum companies like the ONGC and the IOC following a series of sector reforms and deregulation of oil prices between 1990 and 2002 have the highest market capitalization and are listed under India's highly profitable commercially run nine companies Called 'Nava Ratnas' or Nine Gems". However, India's coal and power sector institutions continue to suffer from problems of government micromanagement, lack of enterprise autonomy, unclear lines of operational and regulatory authority and the unclear division of central, regional and state authority.

India's public energy enterprises are ultimately accountable to the Parliament and government ministers exercise specific control over them. Members of Parliament allied with trade unions are able to influence public enterprises in personnel matters and to push social goals. Except for top management posts, public sector energy employees enjoy civil service security of tenure. The power companies officials are government appointees, many of these are from the civil service and others are on contract basis. Many of the top officials of power companies shy away from making tough commercial decisions for fear that they may be transferred and or their contracts will not be renewed. With the government firm in the driver's seat, political concerns rather than commercial interest presently run India's state power sector companies with a few exceptions.

Managed by government appointees and responsible to parliament rather than to their customers, energy sector enterprises have little operational or managerial autonomy. Although laws such as the Electricity Act appear to grant state entities considerable autonomy, in practice these enterprises must obtain approval of state governments and often from the highest political level for most decisions affecting financing, tariffs, borrowing, salaries and personnel actions. State governments guided by their political considerations keep tariffs low and allocate the bulk of their sector investments to power generation. Where some reform-minded state governments such as Andhra Pradesh made efforts to stay the course on tariff reforms, they faced continued political opposition. As power enterprises have no operational autonomy and incentive to manage efficiently1 their financial dependence on government increases. The SEBs inability to pay for coal has exacerbated Coal India's another public enterprise cash flow problems.

Under die energy sectors current institutional setup the roles of various government agencies in policy formulation, operation and regulation are not clearly defined. Regulation is carried out by various departments of the central and slate governments, the CEA and the SEBs themselves. In each of India's five regions, a Regional

Electricity hoards REB: and associated Regional Level Development Commission (RLDC) oversees the operations of the regional grid.

The REBs manage the operation and promote interconnection among each regions constitutive power systems. Central regulation of the power sector however has conflicts between Centre Utilities and SEBs. Tariffs for the NTPC and the bulk supply to the SEBs are centrally set and not mutually agreed upon. States often feel that the CERC favors central utilities over state needs. NTPC has received large injections of government equity, which financed investment that helped it earn satisfactory rates of return.

As observed earlier. India's energy sector treats energy demand forecasts and investment planning as sequential rather than interdependent steps in the planning process. Planning continues to be predicated on demand forecasts extrapolated from consumption trends with alternative scenarios. No attempt is being made to assess the crucial problems of unmet demand and the impact of price on demand. Planning efforts therefore fail to adequately address major constraints particularly the serious shortage of resources.

Investment planning is now generally focused on optimizing the mix of new generation projects, yet where large shortages in capacity exist these are only marginally beneficial. Investment piam11m1g has failed to consider the substantial benefits from efficient operations such as plant rehabilitation and loss reduction, which are far less sensitive to such supply side uncertainties as fluctuations in world oil prices, time availability of natural gas amid construction delays.

Planning arrangements also systematically focus on generation Projects which has resulted in under investment in much needed T&D which in turn jeopardizes least cost operation and encourages higher reserve margins. In addition, the Indian power market (interstate, centre to state, regional and inter-regional has significant problems requiring near-term solutions: over and tinder use, lack of adequate metering, generation shortages, inadequate transmission facilities, poorly documented agreements and regional conflicts of interests.

Restrictive government institutions and government imposed harriers such as tariffs licensing requirements government control of the import sector and foreign exchange controls rarely take commercial considerations into account and impede both energy trade and investment. Even with the power supply contracts determined by the MOP. The current institutional setup through the government-approved fuel allocation schemes, tariff notification and coal linkages with electric utilities impede efficient Operations and investment. For example, because of government-approved gas allocation the main Hazira-Bijaipur—Jagdishpur (HBJ) gas pipeline was not used to capacity.

The institutional structure in the energy sector is in transition since the 1990s with the setting up of the regulatory bodies at the federal and the state levels. CERC at the federal level and SREC at the state level have been set up. These commissions are assigned with the task of regulating tariffs anti promoting competition and efficiency in India's electricity supply industry. Coal and oil sectors continue to be regulated through various government departments which are part of the Ministry of Coal and Ministry of Petroleum.

Review Questions

1. What do you understand by management of energy system and why energy systems management is required?

2. What are the different strategies involved in energy management.

3. Explain the various steps and methodologies involved in energy management system?

4. What is an energy management? Discuss the basic steps of an effective energy management program, and how energy management is linked to sustainable energy development?

5. What is the concept of sustainable development and explain how it can be achieved through energy systems planning and control?

6. What is integrated energy planning and its necessity?

7. What is Energy Management system? Generalize the importance of energy management system in this century?

8. What are the different phases involved in integrated energy planning?

9. Explain in detail the role of different institutions involved in energy management across the country?

Objective type questions

1. The objective of energy management is

a) To minimize energy costs   b) To minimize environmental effects

c) a & b   d)None of the above

2. Which one is the key element for successful Energy Management?

a) Top management support   b) Planning
c) Monitoring   d) Training
3. The objective of energy management is

a) Minimizing energy costs   b) Minimizing wastes
c) Minimizing environmental degradations   d) all of the above

CHAPTER 10

This chapter edifies about the saving of electricity in the various sectors; that can be done by thorough survey and analysis of energy consumption. Now a day's whole world get worried with the environmental problems while combustion or production of energy from various resources. But still these resources plays prominent role in the coming future. Instead of reducing energy production, we can switch to program like saving of energy by surveying the equipments; i.e., finding out energy losses or unnecessary utilization that can be possible by doing energy auditing.

10.1 Energy Audit

An energy audit is an inspection, survey and analysis of energy flows for energy conservation in a building, processor system to reduce the amount of energy input into the system without negatively affecting the output(s).

Energy Audit is the key to a systematic approach for decision-making in the area of energy management. As per the Energy Conservation Act, 2001; Energy Audit is defined as "the verification, monitoring and analysis of use of energy including submission of technical report containing recommendations for improving energy efficiency with cost benefit analysis and an action plan to reduce energy consumption".

Information gathered from the energy audit can be used to introduce energy conservation measures (ECM) or appropriate energy-saving technologies, such as electronic control systems, in the form of retrofits. Energy audits identify economically justified, cost-saving opportunities that result in significantly lowered electrical, natural gas, steam, water and sewer costs.

An energy audit, therefore, is a detailed examination of a facility's energy uses and costs that generates recommendations to reduce those uses and costs by implementing equipment and operational changes. An important part of energy auditing is energy accounting/bill auditing. Energy accounting is a process of collecting, organizing and analyzing energy data.

Creating energy accounting records and performing bill audits can be done internally without hiring outside consulting firms. Also, while energy audits as a whole will identify excessive energy use and cost-effective conservation projects, bill auditing will assist in identifying errors in utility company bills and beneficial rate and service options. It could provide an excellent opportunity to generate savings without any capital investment. In addition, accurate data from energy accounting/bill auditing is crucial.

Role of Energy Auditor:

An energy auditor is an individual who inspects and evaluates home energy efficiency levels. As part of this inspection, the energy auditor will measure energy consumption, track heating and cooling loses, and check the operation and efficiency of heating, ventilation, and air conditioning (HVAC) systems. Based on the results of the audit, he or she will discuss the benefits of potential energy upgrades and improvements with the homeowner.

The home energy auditor inspects the home and documents its energy characteristics, such as construction techniques, insulation levels, window efficiency, wall-to-window ratios, heating and cooling system efficiencies, water heating system efficiency, and solar orientation of the home. Diagnostic testing, such as blower door air tightness testing and duct system leakage testing, are often part of the audit.

The data gathered by the home energy auditor is input into a computer program and a model of the home's energy consumption is generated. Using this model along with local utility rates, estimates of the home's energy costs can be provided. Along with the energy cost analysis, the home owner also receives a report listing cost-effective options for improving the home's energy efficiency.

Here are the specific goals of an energy audit are to:

1. Identify the type, size, condition, and rate of energy consumption for each major energy using device.
2. Recommend appropriate energy conservation, operation, and maintenance procedures.
3. Estimate labor and materials costs for energy retrofits.
4. Project savings expected from energy retrofits.
5. Note current and potential health and safety problems and how they may be affected by proposed changes.
6. Explain behavioral changes that will reduce energy waste.
7. Provide a written record of decision making.

10.2 Energy Audit and energy reporting process

Energy Audit is the key to a systematic approach for decision-making in the area of energy management.

As per the Energy Conservation Act, 2001, Energy Audit is defined as "the verification, monitoring and analysis of use of energy including submission of technical report containing recommendations for improving energy efficiency with cost benefit analysis and an action plan to reduce energy consumption".

Type of Energy Audit

The type of Energy Audit to be performed depends on:
- Function and type of industry
- Depth to which final audit is needed, and
- Potential and magnitude of cost reduction desired

Thus Energy Audit can be classified into the following two types.
- i) Preliminary Audit
- ii) Detailed Audit

Preliminary Energy Audit Methodology

Preliminary energy audit is a relatively quick exercise to:

- Establish energy consumption in the organization
- Estimate the scope for saving
- Identify the most likely (and the easiest areas for attention

- Identify immediate (especially no-/low-cost) improvements/ savings
- Set a 'reference point'
- Identify areas for more detailed study/measurement
- Preliminary energy audit uses existing, or easily obtained data

Detailed Energy Audit Methodology

Detailed energy auditing is carried out in three phases: Phase I, II and III.

      Phase I - Pre Audit Phase

      Phase II - Audit Phase

      Phase III - Post Audit Phase

Phase I – Pre Audit Phase:

Step 1: Plan of Action:

- Plan and organize
- Walk through audit
- Informal interview with Energy manager, production/plant manager

Results:

- Resource planning, Establish/ organize energy audit team
- Organize the equipments and time frame
- Macro data collection
- Assessment of current level operation and practices

Step 2:

Conduct of brief meeting/ awareness programs with all divisional heads

Results:

- Building up cooperation
- Issue questionnaire of each department
- Orientation and awareness creation

Phase II – Audit phase.

Plan of action:

Step 3: Primary data gathering, process flow diagram and energy utility diagram

Results:

- Historic data analysis and data collection
- Prepare flow diagrams
- All service utilities system diagram
- Design operating data and schedule of operation
- Annual energy bill and energy consumption pattern

Step 4: Conduct survey and monitoring

Results:

Measurements: motor survey, insulation and lighting survey with portable instruments for collection of more and accurate data. Confirm and compare operating data with design data

Step 5: Conduct of detailed trials/ experiments for selected energy guzzlers

Results:

Trials/ experiments:

- 24 hours power monitoring
- Load variations trends in pumps, fans and compressor
- Boiler/efficiency trails
- Furnace efficiency trials equipments performance experiments etc.

Step 6: Analysis of energy use

Results:

- Energy and material balance and energy loss/waste analysis

Step7: Identification and development of energy conservation opportunities

Results:

- Identification and consolidation measures
- Conceive, develop and refine ideas
- Review the previous ideas suggested by unit personal
- Review the preview ideas suggested by energy audit if any
- Contact vendors for new/efficient technology

Step 8: Cost best analysis

Results:

- Assess technical feasibility, economic viability and prioritization of options for implementation
- Select the most promising projects
- Prioritize by low, medium, long term measures

Step 9: Reporting and presentation to the top management

Results:

- Documentation report presentation to the top management

Phase III: Post audit phase

Step 10: Implementation and follow up

Results:

- Action plan schedule for implementation
- Follow up and periodic review

Guidelines on Energy Audit (Simplified Version)

1. Introduction

An energy audit is a periodic examination of an energy system to ensure that energy is being used as efficient as possible in a building. This guideline is particular written in simplified version to encourage energy self-auditing by owners or users of smaller buildings. By identifying and minimizing wasted energy through an energy audit, you can achieve the following results:

- Conserve non-renewable energy resources which are gradually running out;
- Protect the environment by burning less fossil fuels, e.g. by reducing power generating requirement, thus lessening carbon dioxide emissions which contribute to global warming; and
- Save energy and reduce running costs.

2. General Self-auditing Procedure

2.1 Collect up-to-date information for the following (Form 1):

- Building details
- Energy bills for electricity, town gas/LPG, etc. for the present and past 2 to 5 years

2.2 Carry out a walk-through of the premises to identify obvious areas of energy wastage and opportunities for energy saving. Form 2 with a detailed checklist is prepared to assist recording results of the survey.

2.3 Implement energy saving opportunities identified in the survey. The opportunities could be implemented with practically no cost implication, e.g. through good housekeeping, or with some capital cost investment, e.g. retrofit fluorescent luminaries with electronic ballasts.

Review Questions

1. What is energy auditing and explain the reporting process of energy audit?

2. Explain the classification of energy auditing in detail?

3. What is the role of energy auditor in industries and explain his importance?

4. Case Study: Do energy auditing in FED department, K.L University, AP and give the ideas to conserve energy?

Form 1

Building Information and Historical Energy Consumption

1   Name of Building: _____

2   Type of Premises: *Office/Shop/Restaurant/Workshop/Warehouse/Residential/Other please specify _____

3   Address: _____

4   Gross Floor Area of Building: _____ m^2

5   Year Built: _____

6   Approximate number of occupants: _____

7   Hours of Operation:

Monday - Friday		Hours/day
Saturday		Hours
Sunday		Hours
Annual Total Hours		Hours/year

8   Records of Energy Bills:

Electricity Consumption (kWh)

Year	Jan	Feb	Mar	Apr	May	Jun	Jul	Aug	Sep	Oct	Nov	Dec	Annual Total

Town gas or Central LPG (Unit)

Year	Jan	Feb	Mar	Apr	May	Jun	Jul	Aug	Sep	Oct	Nov	Dec	Annual Total

Other Fuel (Please specify type & unit used)

Year	Jan	Feb	Mar	Apr	May	Jun	Jul	Aug	Sep	Oct	Nov	Dec	Annual Total

Form 2

Check List for Energy Management Opportunities (EMO)

Energy Management Opportunities	EMO Reference
1. Lighting Equipment:	
a) Lighting turned on unnecessarily	1a
b) Tungsten Filament Lamps	1b
c) 1 switch controlling 2 or more luminaires that are not required to be turned on simultaneously for a task	1c
d) Area that are over-provided with lighting	1d
e) Conventional ballasts used for fluorescent lamps	1e
f)*	1f
g)*	1g
2. Air Conditioning:	
a) Doors or windows are open when air conditioning is operating	2a
b) Temperature setting is unnecessary low for summer and high for winter	2b
c) Air filter is not cleaned regularly	2c
d) Condensation outside air duct	2d
e) Chilled water leakage	2e
f)*	2f
g)*	2g
3. Appliances:	
a) TV is left turned on when the room is vacant	3a
b) Transformers for appliances such as modem, mobile phone chargers to be de-energised when unused	3b
c) Ventilation fan is turned on unnecessarily	3c
d)*	3d
e)*	3e
4. Water	
a) Hot water piping not properly insulated	4a
b) Leakage in shower head or water tap	4b
c) Water heater turned on unnecessarily	4c
d)*	4d
e)*	4e
5. Others*	
a)	5a
b)	5b
c)	5c
d)	5d
e)	5e
f)	5f
g)	5g
h)	5h
i)	5I
j)	5j

* To be filled in by the auditor as necessary

The syllabus as per course hand out:

Lec. No.	C.lev	Learning Objective	Topics to be covered	Reference/ page number
1	C1.1	To Understand what is Energy in different disciplines	Definition of energy	1
2	C2.1	To Understand the different forms of Energy	Diff forms of energy	9
3	C2.1	To Understand Renewable &non-renewable energy forms	Renewable &non-renewable energy	19
4	C2.4	To Understand the Modern forms of Energy	Modern forms of Energy	31
5	C2.1	To Understand the various supply chains of Energy	Supply chains	104
6	C1.2	To identify the different technical aspects of Energy	Technical aspects of energy	41
7	C1.2	To identify the different economic aspects of Energy	Economic aspects of energy	53
8	C1.4	To identify the Environmental aspects of Energy	Environmental aspects of energy	74
9	C1.5	To Understand the institutional aspects of energy	Institutional aspects of energy	112

10	C1.1	To Understand how quality of life is related to energy	Quality of life	153
11	C3.1	To Understand the cost and performance of energy	Cost and performance	67
12	C3.3	To Understand the hidden cost of energy	Hidden cost	64
13	C2.2	To know the definition of energy efficiency	Energy efficiency	125
14	C2.2	To understand the concept of overall efficiency	Overall efficiency	125
15	C4.1	To understand the Application of end use technology at present	End use technology	120
16	C5.1	To understand the concept of energy usage	Energy usage	120
17	C5.1	To understand the application of Energy usage	Energy usage	120
18	C5.4	To know the quality of life in different areas	quality of life in rural and urban areas	153
19	C5.1	To understand the effect of population on energy development	population demographics	139
20	C5.1	To understand the effect of population on energy usage	population demographics	139

21	C5.3	To understand the relationship of energy with economic poverty	economic poverty	**145**
22	C5.3	To understand the concept of economic poverty related to energy	energy poverty	**145**
23	C6.1	To understand the impact of energy on environment	impact of energy on environment	**80**
24	C6.1	To understand the concepts of climate changes and its impacts	concepts of climate changes and its impacts	**85**
25	C6.2	To understand ecological foot prints	Ecological foot prints of an individual	**97**
26	C6.2	To understand ecological foot prints of a an organization	Ecological foot prints of an family, an organization	**97**
27	C6.2	To understand ecological foot prints of a an region	Ecological foot prints of a region	**97**
28	C6.3	To understand the issues of sustainable development	Sustainable development issues	**178**
29	C6.4	To understand the concepts of energy usage	energy usage with respect to sustainable development	**178**
30	C6.4	To analyze the concepts of climate changes and its impacts	energy usage with respect to sustainable development	**178**
31	C7	To understand the concepts of	Energy systems: past, present and	**165-171**

		energy systems	future	
32	C7	To apply the concepts of energy systems	Energy systems: past, present and future	**165-171**
33	C7	To analyze the and the concepts of energy systems	Energy systems: past, present and future	**165-171**
34	C7.1	To understand the concepts of Management of energy systems	Management of energy systems	**187**
35	C7.3	To understand the concepts of climate changes and its impacts	planning and controlling of energy systems	**199**
36	C7.2	To understand the concepts of energy planning	Integrated energy planning	**206**
37	C8.1	To understand the role of institutions in managing energy systems	role of institutions in managing energy systems	**211**
38	C8.2	To understand the the economic and industrial activity effect on energy systems	the economic and industrial activity effect on energy systems	**211**
39	C8.2	To analyze the the economic activity effect on energy systems	the economic activity effect on energy systems	**211**
40	C8.2	To understand the industrial activity effect on energy systems	The industrial activity effect on energy systems	**211**
41	C9.1	To understand the principles of energy audit	Energy audit, principles of energy audit	**218**
42	C9.2	To understand the basics of	basics of energy estimation	**67**

		energy estimation		
43	C9.1	To understand the concepts of reporting the energy audit	energy audit and energy reporting process	221
44	C9.3	To analyze the concepts of energy audit and its report	energy audit and energy reporting process	221
45	C9.4	To take up a case study on the energy audit of an Institution	case study: the energy audit of an institution	221

----:@@@@@ :----

www.ingramcontent.com/pod-product-compliance
Lightning Source LLC
Chambersburg PA
CBHW071133220526
45467CB00015B/919